**The Forensic Digest**
The Official Journal of the International Academy of Forensic Professionals and
The Academy of Forensic Nursing Science
Spring Summer 2012 

The Forensic Digest
Spring Summer 2012

# CONTENTS

16 From the Editor's Pen
• Looking to Tomorrow's Leaders

POLICE

# FEATURES

The identity of Jack the Ripper, often regarded as one of the world's most notorious serial killers, has remained a topic of great debate for over a century. As one of the most popular and controversial serial killers of all time, the unknown murderer who terrorized London's Whitechapel area between August and November 1888 has baffled countless researchers, historians, law enforcement officials, and investigators for over a hundred years. Although many individuals have presented evidence claiming to reveal the true identity of the killer, no single, concrete conclusion exists.

Today, Native Americans face many challenges in their movement towards self-determination. Native Americans are overrepresented as offenders and victims of crimes in our criminal justice system. Native Americans are struggling to overcome political and economic marginalization. Discrimination and stereotyping of Native Americans is sewn into the fabric our Western culture. Alcoholism, poverty, illiteracy, and domestic violence plague the lives of Native Americans on and off the reservations. Indigenous values and movements toward self-determination often conflict with current criminal justice systems.

It has become more important than ever for modern police officers to recognize their responses to stress, what to do with those responses, and the appropriate methods of verbalizing their feelings. Subsequent to a critical incident, officers may experience a wide range of emotions, including fear, anxiety and confusion.

# International Academy of Forensic Professionals

*Globally Representing All Forensic Disciplines*

**The International Academy of Forensic Professionals (IAFP)** is a member-based global institute of higher learning, networking and mentoring representing all disciplines of forensic science. We attract and connect forensic professionals through cutting-edge scientific education, events and interactive media. We uphold the core values of knowledge, integrity, competency, and best practices.

## MEMBER CATEGORIES

### *Professional Member*

An individual professionally certified, licensed, or directly involved in the forensic disciplines such as fire investigations, social workers, registered nurses, law enforcement, researchers, psychologists, physicians, EMTs, educators, coroners, and death investigations, etc. Annual membership entitles the holder to all of the benefits and privileges granted by IAFP. Membership is valid for one year.

**Membership Dues: $67 U.S.**

### *Group Member*

Educational Institutions, law enforcement, hospitals, vendors and other organizations that would like to officially align and support the IAFP mission. Membership entitles the holder to all of the benefits and privileges granted by IAFP. Membership is valid for one year. *Contact corporate office for group membership information.*

### *Lifetime Membership*

Lifetime membership is valid for the duration of the present lifetime of the holder. Life membership shall continue until the member's death or until the member relinquishes membership.

Lifetime Membership entitles the holder to all of the benefits and privileges granted by IAFP.

**Membership Dues: $1500.00 U.S.**

### *Honorary Member*

Honorary membership is reserved for individuals, members or non-members, past or present, who have contributed distinguished service to the field of forensic science. Eligibility for Honorary Member status shall be based upon nomination by majority vote of the Board of Directors and confirmation by majority vote. Membership entitles the holder to all of the benefits and privileges granted by IAFP. Honorary membership shall continue until the member's death or until the member relinquishes membership.

### *Founding Member*

Special status of Founding Member is granted to members who are responsible for the initial creation of the International Academy of Forensic Professionals. This special status is valid for the lifetime of the member. Founding Members are eligible for all benefits and privileges of IAFP. There are three founding members. This membership is no longer available.

## MEMBER BENEFITS

The ***International Academy of Forensic Professionals***, multidisciplinary in scope, offers membership to individuals working within the forensic science arena. Member professionals must be in good standing with the licensing and certification boards in their jurisdiction of practice. The Board may decline membership without assigning any reason. Memberships are valid for one year. All member fees are non-refundable and non-transferable. Professional members may vote and hold office.

- An original certificate of membership and accompanying wallet sized card reflecting date of membership
- Complimentary access to the official E-journal, Forensic Digest, featuring cutting-edge articles from leading forensic professionals; published twice annually
- Complimentary forensic science courses with continuing education hours
- E-journal advertising for members at reduced rates
- Access to the Forensic Science Portal Information Resource Center, providing breaking news, research, cutting-edge technology, forensic science case histories and current publications
- Educational conferences and events at reduced rates
- Job listing information for forensic professionals
- Opportunity to develop forensic science courses
- Opportunity to publish in the Journal and Newsletter
- Participate in webinars
- Network with internationally recognized forensic professionals
- Members Directory with access to other member colleagues
- Arson consultation, investigation and profiling services

**INTERNATIONAL ACADEMY OF FORENSIC PROFESSIONALS**

**255 North El Cielo Road, Suite # 140-195**
**Palm Springs, CA 92262**
**Telephone: 760.322.9925 * Fax: 760.322.9914 * info@tiafp.com**
**WWW.TIAFP.ORG**

AMERICAN
INSTITUTE
OF
FORENSIC
EDUCATION

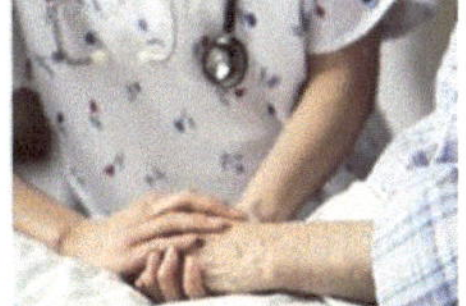

**American Institute of Forensic Education, Inc.**
255 N. El Cielo Road #140-195, Palm Springs, CA 92262 Telephone 760.322.9925
Fax 760.322.9914 Email: info@taife.com **Website: www.taife.com**

## MEMBER CATEGORIES

**The Academy of Forensic Nursing Science (AFNS)** offers membership to those registered nurses interested or engaged in the field of forensic science. The licensed professionals must be in good standing with the licensing boards in their jurisdiction of practice. The Board may decline membership without assigning any reason. Professional, and group memberships are valid for one year. All member fees are non-refundable and non-transferable. Professional members may vote and hold office.

### *Professional Member*

A professionally licensed individual, interested or directly involved in a forensic discipline. To register as a professional member, a license as a registered nurse is required.
Annual Fee: $65 US

### *Group Member*

Educational institutions, law enforcement, hospitals, vendors and other organizations that would like to officially align and support the AFNS mission. Membership entitles the bearer to all of the benefits and privileges granted by AFNS.
*Contact the membership office for more information on fees.*

### *Lifetime Member*

Lifetime membership is valid for the duration of the present lifetime of the holder. Life membership will continue until the member's death or until the member relinquishes membership. Lifetime Membership entitles the holder to all of the benefits and privileges granted by AFNS.
Membership Fee: $1500 US

### *Honorary Member*

Honorary membership is reserved for individuals, members or non-members, past or present, who have contributed distinguished service to the field of forensic science. Eligibility for Honorary Member status shall be based upon nomination by majority vote of the Board of Directors and confirmation by majority vote. Membership entitles the holder to all of the benefits and privileges granted by AFNS. Honorary membership shall continue until the member's death or until the person relinquishes membership.

### *Founding Member*

Special status of founding member is granted to members who are responsible for the initial creation of the Academy of Forensic Nursing Science. This special status is valid for the lifetime of the member. Founding Members are eligible for all benefits and privileges of AFNS as announced, including a Founding Member lifetime pin and membership card. There are four founding members. This membership is no longer available.

## MEMBER BENEFITS

- An original certificate of membership and accompanying wallet sized card reflecting date of membership
- Complimentary access to the official AFNS E-journal, Forensic Digest, featuring cutting-edge articles from leading forensic professionals; published twice annually
- Complimentary Members-only E-magazine, the Forensic Times, focusing on news, articles and views from member colleagues; published twice annually
- Access to the Forensic Science Portal Information Resource Center, providing breaking news, research, cutting-edge technology, forensic science case histories and current publications
- Complimentary members-only newsletter, the Forensic Voice, providing up dates and alerts
- Extensive online self-paced accredited continuing education forensic science courses
- Access to mentors and experienced forensic professionals who provide advice, support, and knowledge to those members less experienced
- Job listings and career development information for nursing professionals at every stage in their careers
- Access to educational conferences and events at reduced rates
- Opportunity to develop forensic science courses
- Opportunity to publish in the journal and magazine
- Participate in webinars
- Communicate with internationally recognized forensic professionals
- Members Directory with other member colleagues

**THE ACADEMY OF FORENSIC NURSING SCIENCE**

**255 North El Cielo Road, Suite # 140-195Palm Springs, CA 92262**
**Telephone: 760.322.9925 * Fax: 760.322.9914 * info@tafns.com**



FOR MEMBERSHIP INFORMATION VISIT WWW.TAFNS.COM OR TELEPHONE 760.322.9925

# THE FORENSIC DIGEST

The Official Journal of the International Academy of Forensic Professionals and the Academy of Forensic Nursing Science





*The Forensic Digest*, published semi-annually, is the official journal of the International Academy of Forensic Professionals (IAFP) and the Academy of Forensic Nursing Science (AFNS). This on-line journal, devoted to forensic practitioners of all disciplines, represents the professional work of those engaged in forensic science at all levels.

We recognize that practitioners in this area of specialization are many and varied. Our goals are far-reaching and reflect a commitment to provide thoughtful and balanced articles, opinions and research to our readers including forensic case studies, interviews with forensic professionals, current news, educational events, advertising and other cutting-edge forensic science information.

We encourage first time authors and experienced writers to submit articles for publication and invite those interested in doing so to consult our submission guidelines at www.tiafp.org. This is a journal for all of us. Join us in sharing knowledge, opinions, experience and ideas.

The Journal reflects our recognition of the many disciplines involved in forensic practice and is dedicated to promoting the continued advancement of forensic science. The editorial staff is committed to the journalistic ideals of excellence.

Please enjoy this edition of the Forensic Digest and do not hesitate to share your views and opinions with us!

## PAST ISSUES:

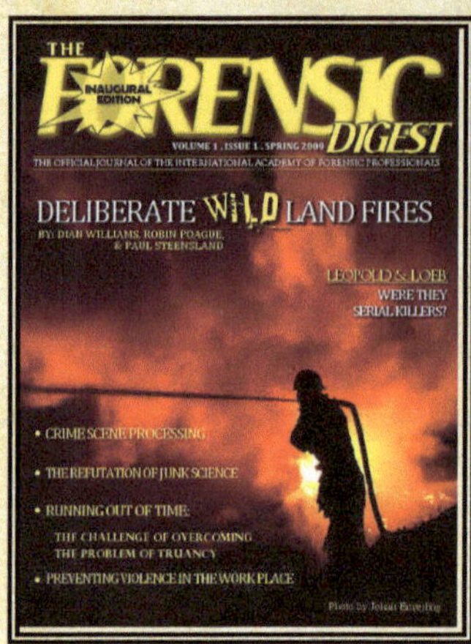


2009

2010

2011

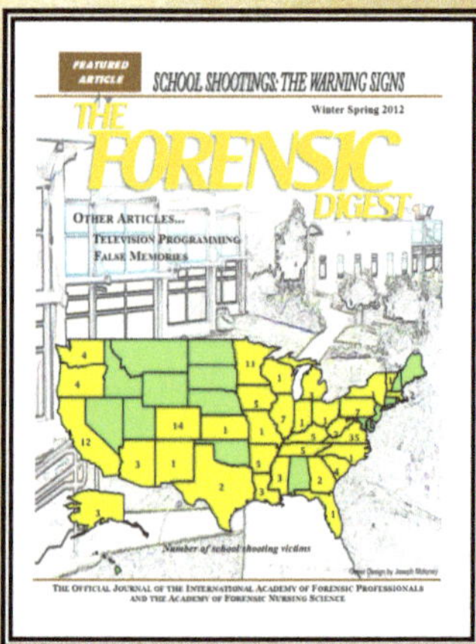


2012

| SUBSCRIBE! | ADVERTISE! |
|---|---|
| **RATES** | **FREQUENCY** |
| **Full Page** **$199** | 3x |
| **Half Page** **$99** | 3x |
| **Quarter Page** **$49** | 3x |

 FOR SUBSCRIPTION INFORMATION VISIT WWW.THEFORENSICDIGEST.COM OR TELEPHONE 760.322.9925

Center for Arson Research is an internationally recognized consulting company that provides education on fire setting behavior in youth and adults. Experts at the Center conduct face-to-face evaluations of known or suspected fire setters, fire bombers and domestic terrorists. Dr. Dian Williams, president of the Center for Arson Research, is a qualified expert witness in the motivation and risk assessment of fire setting behavior. She also participates in arson task force meetings around the country and abroad.

The second edition of her textbook,

*Understanding the Arsonist from Assessment to Confession*

is due for completion this year.

To reach the Center for Arson Research

Telephone: 215.843.2115 or

Email: drarson@msn.com

# Looking to Tomorrow's LEADERS

This edition of the Forensic Digest features a column co-authored by the publisher, Faye Battiste-Otto and the editor-in-chief, Dian Williams.

Over the past several years, the writers have had ongoing discussions about the concept of leadership. What distinguishes a leader from others working in the same field? One element of leadership is that of the power to influence others (hopefully for the common good). A leader must have the skill to induce people to follow him/her towards established goals.

An impressive study by House, Hanges, Javidan, Dorfman & Gupta (2004) studied the attributes of leaders in 62 countries. As a result, they were able to identify positive and negative qualities found in leadership. Positive leaders were described as: motivational, dynamic, trustworthy, informed, good communicators, and honest, among other attributes while negative leaders were described as asocial, ruthless, ego-centered and irritable. Both the positive and negative qualities were universally agreed upon by the 62 countries involved in the research project.

Leaders apparently are both born and made. Conger (2004) believes that leaders are born with certain skills that are developed over time. Early research into leadership skills revolved around the notion that certain individuals had traits that distinguished them from others and in effect, helped them to be successful. Stodgill (1974) believed that effective leaders have certain identified traits such as the ability to motivate others, emotional control , self-confidence and flexibility. Bass (1990) looked at three categories of leadership traits he labeled as: intelligence, ability and personality.

***...a key attribute of a successful leader is that of emotional intelligence which is the ability to insightfully understand others*** — *Goleman (1995)*

Leaders are agents of change who have a long-range vision that focuses on the goals and mission of an organization and how to best accomplish them from short and long-term perspectives. One of the questions for organizations is whether those skills and traits can be learned. If we philosophically agree that people can develop leadership traits they do not innately have, are they ever too old to learn them? This is a vitally important question because as the leaders of today (baby boomers) are ready to retire, where are the leaders of tomorrow in forensic nursing, policing, law, death investigation, social services, psychology and the myriad other disciplines that comprise the world of forensic practice?

Goleman (1995) believes that a key attribute of a successful leader is that of emotional intelligence which he describes as the ability to insightfully understand others. According to Goleman, emotional intelligence is comprised of a number of elements, the most important of which he believes s empathy as a skill for managing and leading others.

Goleman's emotional intelligence variables include among others:

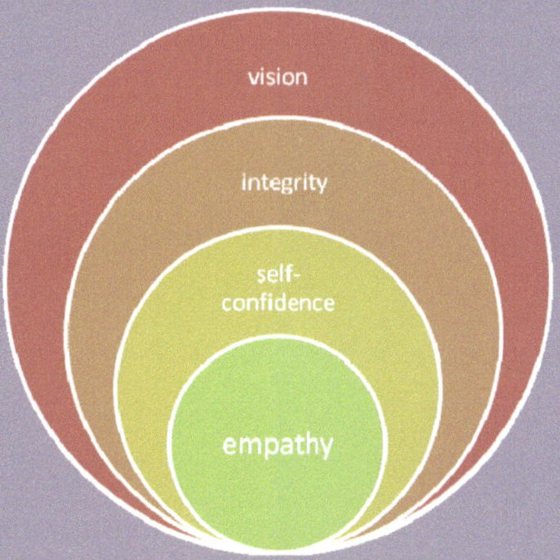


There is a major body of work in nursing on patterns of knowing, developed by Barbara Carper in 1978. Dr. Carper proposed that certain diverse patterns of knowledge formed the basis for understanding the practice of nursing as both art and science. The writers suggest that the principals involved in patterns of knowing can be adapted to a leadership theory, as well. According to Carper (1978) there are four ways of knowing: ethically, empirically, aesthetically and personally. We suggest that positive leaders also possess knowledge in those four domains.

Empirical knowledge is that which is gained through observation and experience. Such information is measurable (i.e. this has worked 9 of every 10 times) fact-based and objective (Fawcett, Watson, Neuman, Walker, & Fitzpatrick, 2001). Ethical knowledge involves following the guidelines and standards of our respective disciplines and in doing what is right even when no one is watching. Because forensic practitioners frequently interface with the legal and mental health systems, strong leaders must feel confident in knowing the ethical thing to do, even if faced with controversy and/or high profile cases. Personal knowledge implies that forensic leaders recognize who they are and what they believe and value. Finally a leader understands that there is an art as well as a science in connecting to and communicating effectively with others to achieve goals (Fawcett, et al., 2001).

As with other studies, Bennis& Goldman (1997) identified some common qualities of a leader, a number of which are found below:

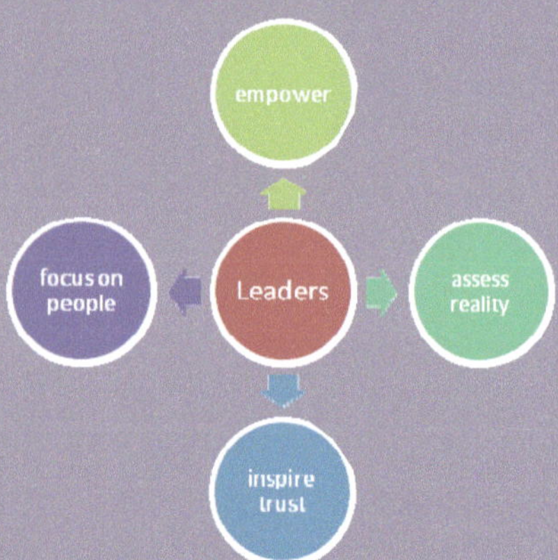


One aspect of leadership that is understudied is that of the skill involved indecision-making. We know that leaders must make decisions every day, some of minor significance and some that may profoundly affect the organization and the individuals who work there. Critical decisions are seldom made with complete knowledge of every factor, and all possible alternatives and consequences. According to McConnell (2000) successful decision-making depends on the judgment, experience and knowledge of the person in charge. A leader with poor relational skills, personality traits that do not foster trust along with lack of experience and poor adaptability may face obstacles related to successful decision making (Sullivan & Decker, 2009).

We have probably all worked at one time or another for and with leaders who we either admired or found extremely frustrating and/or aggravating. We tend to remain longer in a job when we believe in the leader and agree with her/his style. However, the most common styles are not necessarily the most creative, effective or inspirational. We invite the reader to think of positions you have held where the following styles were utilized and how well you think they worked:

a. *autocratic* – the person issues orders and expects them to be carried out. This type of leader believes that employees are at their best when they are told what to do

b. *bureaucratic-* this leader relies on policies, rules and regulations and thinks employees might take advantage of the employer if rules are flexible

c. *laissez-faire* – the leader believes employees do best if left on their own to figure things out. Under this style, employees sometimes feel they have no guidance at all

As we consider how to advance the practice of the ever-growing specialty areas found in forensics, it seems important to help the managers of today develop the skills, attributes and knowledge they need to become the leaders of tomorrow. On our part, Faye and I will continue our dialogue about how to educate and mentor our future forensic leaders.

Suggestions anyone?

---

**References :**


Bass, B. (1990). *Bass and Stodgill's handbook of leadership: Theory, research and managerial applications* (3rd ed.). New York: Free Press.

Bennis, W. & Goldsmith, J. (1997). *Learning to lead: A workbook on becoming a leader.* Reading, MA: Perseus Books, pp. 4, 9-10.

Carper, B. (1978).Fundamental patterns of knowing in nursing. *Advances in Nursing Science*, 1(1), 13-23.

Conger, J.A. (2004). Developing leadership capability: What's inside the black box? *Academy of Management Executives*, 18 (6), 136-139.

Fawcett, J., Watson, J., Neuman, B., Walker, P.H. & Fitzpatrick, J. (2001). On nursing theories and evidence. *Journal of Nursing Scholarship*, 33(2), 115-119.

Goleman, D. (1995). *Emotional intelligence*. New York: Bantam.

House, R., Hanges, P.J., Javidan, M, Dorfman, P. & Gupta, V. (Eds.). (2004). Culture, leadership and organizations. *The GLOBE Study of 62 Societies.* Los Angeles: Sage.

McConnell, C.R. (2000). The anatomy of a decision.*Health Care Manager*, 18(4), 64-73.

Stodgill, R.M. (1974). *Handbook of leadership*. New York: Free Press.

Sullivan, E. & Decker, P.J. (2009).*Effective leadership and management in nursing.* Upper Saddle River, NJ: Pearson/Prentice Hall

# Serial Arson Investigators Seminar

**Hosted by: NVIAAI**

**Sponsored by: The Nevada and Los Angeles 'High Intensity Drug Trafficking Areas.'**

**Location:** New York New York Hotel& Casino in Las Vegas, Nevada, on the Las Vegas Strip.
**Date:** Monday, May 20th, 2013 to Thursday, May 23rd, 2013 **Time:** 0800-1700 Hrs

**A first of its kind four day seminar on all aspects of serial arson investigation by the investigators and prosecutors who solved these major cases.**

**Case studies of these major investigations will be presented:**

-John Orr- the most significant serial arsonist ever identified
-Paul Keller- Seattle Serial Arsonist
-Thomas Sweatt- DC Serial Arsonist
-Texas Church Arson Series
-Ben Cunha Wildland Series (California)
-Alabama Church Arson Series
-Serial Arson Case Management, ATF NRT

**Mormon Temples Arsons**
**-North Hollywood Serial Arsonist(2011) ***
**-Hollywood Serial Arsonist (2012)***

## Additional Subjects:

**-Dr. Dian Williams will present the profileofan arsonist by sub-type and describe which subtypes produce serial arsonists**
**-Professor Matt Hinds-Aldrich will present his studies on firefighter arsonists**
**-Authors of several arson and serial arson related books to be present**
***pending official adjudication of the case**

**Co-Organizers: SA Dan Heenan, ATF; and Det. Ed Nordskog, L.A. Sheriff's**
**Registration Info:**

**Cost for tuition: $200 for sign-ups prior to December 1st, 2012; $225 for sign-ups after December 1st.**
**Class limited to 300 students; Register early as this event will likely sell out.**

**To register, go to www.lahidtatraining.org.**
**This conference is a MUST for anyone involved in fire investigation**

# CHALLENGES FACING AMERICAN INDIAN CULTURE

*By Charles Cernat II*

# Abstract

**Prior** to European contact, Native Americans had complete control of their own cultures, societal structures, economic systems, and laws. Justice was part of day to day living for Native Americans and when one's behavior disrupted the community, remedial action such as restoration of harmony and healing approaches were implemented by the community.

Upon the arrival of the Europeans, technology was introduced to Native Americans resulting in permanent changes to the economies of many Native groups. Ideologies based on social Darwinism and paternalism were used to justify the treatment of Native Americans. Europeans introduced laws forbidding the practice of Native beliefs in an effort to force Native Americans into their culture.

**Today,** Native Americans face many challenges in their movement towards self-determination. Native Americans are overrepresented as offenders and victims of crimes in our criminal justice system. Native Americans are struggling to overcome political and economic marginalization. Discrimination and stereotyping of Native Americans is sewn into the fabric our Western culture.

Alcoholism, poverty, illiteracy, and domestic violence plague the lives of Native Americans on and off the reservations. Indigenous values and movements toward self-determination often conflict with current criminal justice systems.

## Introduction

**Native American Indians** face many obstacles as they attempt to move toward self-determination. Europeans and Americans used ideologies based on social Darwinism after 1776 to justify the treatment of Native Americans and as Trigger (1985) describes, "It offered a comfortable explanation for the primitive condition of the American Indian and his stubborn refusal to accept the benefits of civilization. White Americans could not be blamed for the tragic failure of natural selection over the course of millennia to produce native North Americans who were biologically able to withstand the impact of Western civilization" (p. 16).

**Today,**

**the same colonial ideologies continue but are called "racism" and these biases affects how other members of society interact with Native Americans *(Mihesuah, 1993).***

This may contribute to the discriminatory treatment of Native Americans in our criminal justice system. According to Nielsen (2009), "Relations between the dominate society and Native Americans are still colored by racism and greed among members of the dominate society, but neglect is also a power force, with many issues still needing resolution" (p. 9). Native Americans are faced with a wide range of controversial issues such as political rights, economic development, environmental crimes, religious freedom, and education. Government policies that prohibit Native Americans from making decisions about the use of their resources,

funds, land, and the ability to provide services for themselves complicates the issues even further.

As Utter (2001) points out, Indian nations are "domestic dependent nations" that have the right to be protected by the federal government as "distinct political communities" (p. 264). According to Wilkins and Lomawaima (2001), "a sovereign nation defines itself and its citizens, exercises self-government and the right to treat with other nations, applies its jurisdiction over the internal legal affairs of its citizens and subparts, claims political jurisdiction over the lands within its borders, and may define certain rights that are inherent in its citizens" (p. 4). However, as Utter (2001) points out,...

**"sovereignty means many things to many people"** (p. 264).

# Native Americans and Gaming

According to the U.S. Census Bureau, an estimated 4 million people list themselves as being Indian or Alaskan Native (2005). This is approximately 1.5 percent of the total U. S. population. Approximately 25.1 percent of Native Americans live below the poverty line and these percentages are often higher on reservations (U.S. Census, 2007:16). Approximately 14 percent of Native Americans are unemployed and percentages on Indian reservations are estimated as high as 40-50 percent unemployed (U.S. Census, 2006).

As federal support for tribal activities diminished, Native Americans have led the movements of gaming as a source of revenue. The development of Indian tribal gaming began in late 1970s and early 1980s (Ross & Gould, 2008). Tribes in Wisconsin, Connecticut, California, and Florida open low stakes bingo halls on their reservation and eventually expanded their gaming enterprises (Ross & Gould, 2008). In 1988, Congress passed the Indian Gaming Regulatory Act to provide a statutory basis for the Industry (Ross & Gould, 2008). According to the National Indian Gaming Association (2003), gross annual revenues now exceed over $14 billion.

As Native Americans expanded their gaming enterprises, an economic conflict developed. Who should control or share in the financial benefits resulting from the money wagered and lost in games operated by Indian tribes? (Mason, 2000).These ideological conflicts centered on questions regarding federalism, states' rights, and Indian sovereignty. According to Mason (2000), "Indian policy has often been driven by conflict over who controls Indian Country-the federal government, state governments, or the tribes themselves" (p. 46).

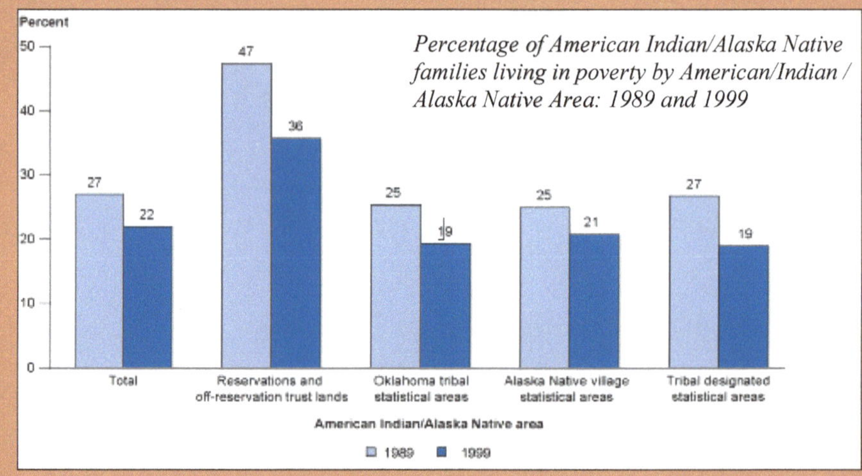


*Percentage of American Indian/Alaska Native families living in poverty by American/Indian / Alaska Native Area: 1989 and 1999*

Opponents of Indian gaming believe that casinos are magnets for crime (Fairbanks, 2003; McCue, 2003). There is often concern about increased court, police, and jail costs especially in adjacent non-Indian communities (Sutherly, 2003; Muir, 2003). These opponents of Indian gaming are also concerned about increases in crime indirectly related to gambling addiction, such as theft and burglary (Huges, 2003). According to Ross and Gould (2006), "In nearly all cases, opposition to Indian gaming has two characteristics: a non-Indian origin and an assertion, usually stated without supporting evidence, that there is a linkage between Indian casinos and criminal behavior" (p. 181). It should be pointed out, however, that not all Native Americans support gaming. For example, the Navajo Nation receives the most publicity for its opposition to gaming and the fear that increased crime will follow casino gaming on their reservation (Henderson and Russel, 1997).

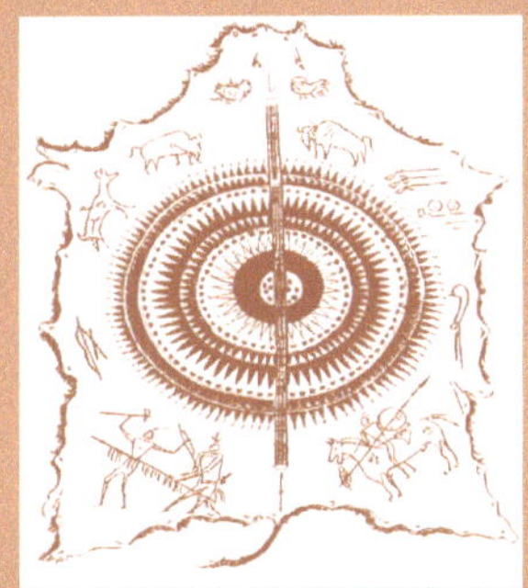

Proponents of Indian gaming dispute linkages between casinos and increased criminal behavior. For example, the National Indian Gaming Association (2003) argues that gaming generally improves the situation of Indians and non-Indians who live in and around Indian casinos. This claim, however, is unsupported. Ross and Gould (2006) state, "It is a common assumption that casinos reduce crime because increases in wages reduce crime"(p. 183). For example, if casinos provide more jobs to low-skilled workers, the rate of crime should decrease and casinos should also reduce crime indirectly by stimulating economic development. However, as Grinols and Mustard (2001) point out, "Indian gaming may harm economic development and raise the rate of crime by undermining the local business climate because casinos may attract an unsavory clientele and encourage prostitution, drug trafficking, and other illegal activities. Although a casino may improve local economies overall, the large increase in pawn shops that accompany new casinos suggest the economic well - being of everyone within surrounding communities may be uneven" (p. 11).

*Gambling has ruined countless lives. The level of crime, suicide, and bankruptcy in a community invariably rises when a casino opens its doors"*

*2001-U. S. Representative Frank R. Wolf*

is unclear if gaming actually benefits Native Americans in their movement toward self determination. As Light and Rand (2005) point out, "Despite mixed results of studies exploring the link between gambling and crime, the perceived connection between casinos and crime is a powerful influence on policymaking; the threat of increased street crime almost invariably is raised in opposition to opening a casino" (p. 97).

The National Indian Gaming Association, however, has a very different perspective.

*According to National Indian Gaming Chairperson Ernest Stevens Jr., "Indian gaming offers hope for the future" (NIGA, 2001). In a report issued by the Wisconsin Policy Research Institute, the Oneida Nation of Wisconsin is also enjoying gaming success. The report states, "The Oneida Tribe .is enjoying its first generation of prosperity in more than two centuries. For the Oneidas, the gaming franchise has been more successful than all previous anti-poverty programs in providing jobs, self-esteem, and a bright future" (Alesch, 1997). According to the Capital Times, "The poverty Rate among Wisconsin Oneidas dropped ten-fold between 1990 and 2000 from 50 percent to 5 percent, in part due to the tribe's casinos" (Nowlen, 2004).*

It should be noted, however, that much of the available research regarding Native American casinos suffers from inconsistencies, methodological problems, weak correlations, and questionable statistical analysis Walker, 2001). However, the conflict and interest in Native American gaming inconsistent with Schattschneider's classic "scope of conflict" arrangement wherein parties with an economic

*http://www.mcclatchydc.com/2012/07/05/155126/in-a-new-twist-indian-tribes-are.html*

interest in an issue seek the arena of government most likely to award them victory (Schattschneider, 1975). According to Peroff (2006), "The biggest underlying reason for the absence of comprehensive research on the relationships between casinos and crime on reservations is that Indian tribes are sovereign nations and are reluctant to approve research by non- tribal members on their reservations for fear of any harmful linkage between casinos and crime" (p. 185).

Proponents and opponents of Indian gaming will likely continue their debate for many years to come. There simply is not enough reliable research data to base a solid argument regarding the positive or negative effects of Indian gaming. It is clear however, that Indian gaming alone has not solved domestic violence, alcoholism, poverty, and illiteracy that plague life on reservations.

# Native Americans and Alcohol Use

*A man drinks a beer while standing with other American Indians on the streets of Whiteclay, Neb. (File photo by William Lauer, courtesy of AP.*

According to Ross and Gould (2006), "The rate of alcohol-related deaths for Native Americans in the United States is seven times that of the general population. For those Native American youth between the ages of fifteen and twenty-four, the alcohol-related death rate is more than twelve times that of the comparable general population, while those Native Americans between the ages of twenty-five and thirty-four, the rate is thirteen times greater" (p. 87). As Grim (2002) points out, substance abuse among Native Americans is a major leading contributor to health problems.

How does one even attempt to explain the rate of alcoholism and levels of individual consumption among indigenous peoples within the United States? According to Gould (2006), "A variety of theoretical propositions, ranging from the biological (indigenous people are genetically different) to the psychological indigenous

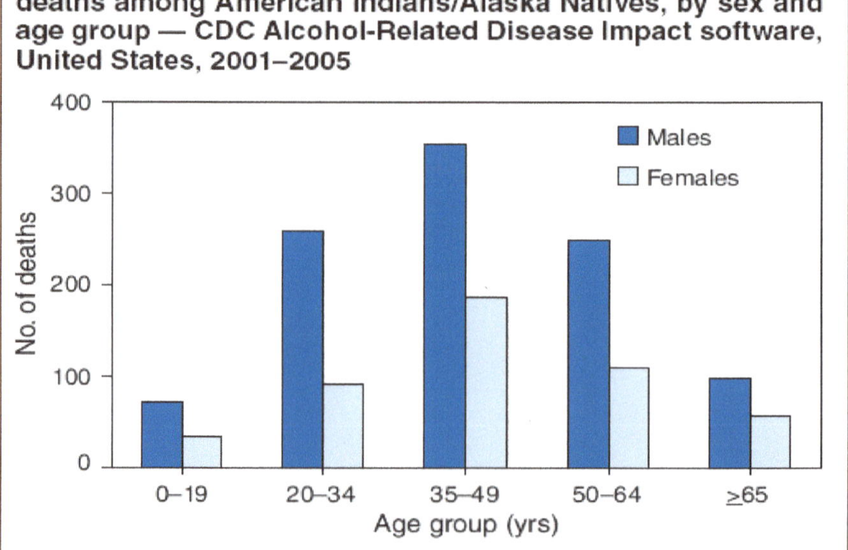


National Center for Education Statistics/Status and Trends in the Education of American Indians and Alaska Natives 2008

**Alcoholism among indigenous people has been found to disorganization, feelings of alienation, and anomie .**

-(Kahn, 1992; Kraus and Butler, 1979)-

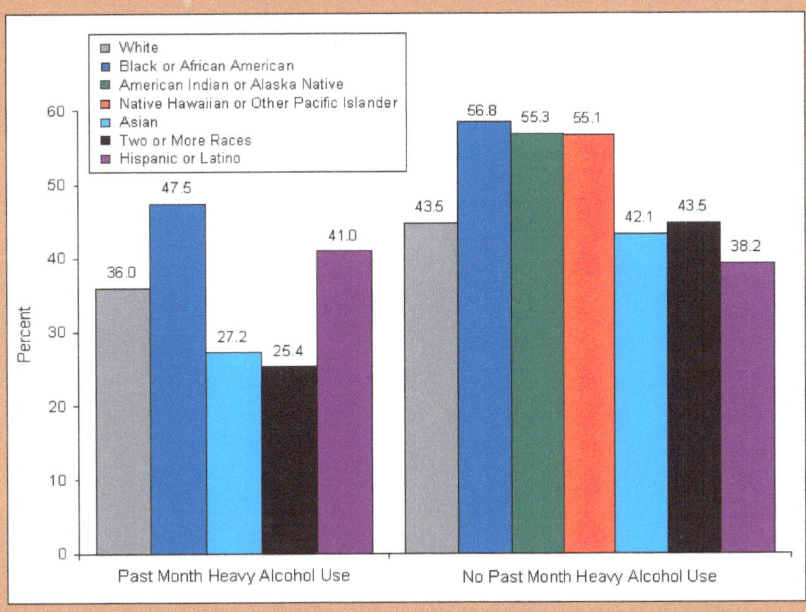


people drink because of low self-esteem, anxiety, frustration, boredom, powerlessness, peer pressure, and isolation) to the socio-culture or environmental (enculturation, governmental paternalism, deprivation, persistence of traditional patterns, poverty, and recreation) are used in an attempt to fully explain high consumption rates among Native Americans" (p. 88). The use of these theoretical propositions to explain the rate of alcoholism among indigenous people varies dramatically (Mall, 1984). Gould (2006) argues, that singly, none of these propositions fully explain alcohol use among indigenous people. However, when one examines these propositions used in combinations, a better understanding of the issue of alcohol use among indigenous people is possible (Gould, 2006).

Studies regarding the impact of alcohol use among indigenous people leave one to question whether indigenous movements towards self-determination are even possible. Alcoholism among indigenous people has been found to intensify social disorganization, feelings of alienation, and anomie (Kahn, 1992; Kraus and Butler, 1979). In addition, alcoholism among indigenous peoples has been associated with increases in violent behavior (Greenfield and Smith, 1999; Reina, 2000). According to Saggers and Gray (1998), alcohol abuse either directly or indirectly, has resulted in increased levels of homicide, health problems, crime, sexual assault, problems with children and a lower quality of life among indigenous people.

Will stereotyping indigenous people as alcoholics and drunks make it easier for some segments of the population to dismiss them? (Mihesuah, 1996). As Gould (1999)points out, drunkenness among the upper and middle income people is less visible than that in the lower-income groups in the United States. A very large portion of the indigenous population have much lower incomes and when they drink, are forced to frequent cheap bars where they are more likely to become easy targets for public criticism and victims of crimes such as theft, mugging, and assault (Gould, 1999). According to Ross and Gould (2006), "The 'out-of sight, out-of-mind' philosophy is so ingrained that excessive drinking only becomes an issue if it directly impinges on the daily lives of whites. There seems to be little concern that beverage outlets are making tremendous profits through the sale of alcohol to indigenous people with drinking problems or, worse yet, to people who were already visibly intoxicated" (p. 95). According to a quote from Benjamin Franklin, "If it be the

design of Providence to extirpate these savages in order to make room for cultivators of the earth, it seems not improbable that rum may be the appointed means (Diamond,1992:1). Are there similarities in how the powerful perceive indigenous people today?

Native American poverty in the US

http://weknowwhatsup.blogspot.com/2010/12/obamas-fiscal-commission-republicans.html

## Native American Offenders and Victims of Crime

*Native Americans comprise about 16 percent of all offenders who enter federal correctional facilities annually (Perry, 2004).*

It is difficult to estimate the exact crime rates for Native Americans. Many of the FBI's Uniform Crime Reports (UCRs) missed rural areas including American Indian Nations where a large number of Native American live (Silverman, 1996). There is also missing information in the Census such as how to classify oneself if you are of mixed raced as is the cases for many Native Americans. With those statistical shootings in mind, however, there are a significant number of Native Americans who are incarcerated when one compares the arrest rates to whites and African Americans. According to Silverman (2009), "Arrest rates from 1984 to 2005 were 4,652 per 100,000 for all Americans. This calculates to 3,975 (per 100,000) arrests for whites, 6,340 for Native Americans and 11,637 for African Americans. The crime rate pattern for Native Americans was higher than whites, but lower than for African Americans" (p. 78). In areas of alcohol-related offenses (liquor violations, drunkenness, and driving while intoxicated), Native Americans had the highest rates when compared to African Americans and whites. Native Americans had a rate of 2.4 times that of the average for all Americans in regards to alcohol related of fense(Silverman,2009).

*Circles- an alternative approach to sentencing offenders in the Yukon. http://www.shantithakur.com/circles.html*

Native Americans as victims of crime depend on the ability of tribal police to arrest perpetrators of violent crimes. According to Luna-Firebaugh, **"Non-Indian perpetrators, even if**

**they live on the reservation or are in a personal or business relationship with an Indian, may only be detained or left to the state or federal police to be apprehended. This can cause lawlessness on a reservation that can't be addressed by tribal police"**(Luna-Firebaugh, 2007:14).

The rates of sexual victimization for Native American women are higher than for any other United States ethnic group (Tjaden and Thoennes, 2006). In addition, rapes involving Native American women are often more severe than rapes committed against other women (Bachman,2004). According to Perry (2004), a distinctive feature of sexual crimes against Native American women is that most perpetrators are non-Indian (86 percent). The high rate of victimization by non-Indians is particularly problematic, because reservation police have no jurisdiction over non-Indians, even on Indian land, and so can only refer these crimes to federal, or (in some cases) state authorities, which often impedes arrest and prosecution of these cases" -Hamby, 2009 (p. 62). Most rape victims, regardless of ethnicity, do not report their rape to the police (Tjaden and Thoennes, 2006). The relationship among tribal, state, and federal laws create issues that often are overlooked. When assaults are committed on reservation land, and the perpetrator is non-Indian, jurisdictional problems arise because reservation authorities are often reluctant to get involved in all but the most severe reservation crimes (Amnesty International, 2007).

**According to Hamby (2006), Native Americans who are victims of sexual assault often do not receive needed services because they are not readily available due to poverty. According to the United States Commission of Civil Rights (2003), "Native Americans receive less funding per capita than any other group for which the federal government has health care responsibilities, including Medical/Medicare recipients, veterans, and prisoners" (p. 13).**

## Community Policing Approach of Proactive Peacekeeping

Bob Thomas, a Cherokee University of Arizona professor, said in a conversation many years ago, "Indians have relationships, Americans have roles" (Luna-Firebaugh, 2007). As Luna- Firebaugh (2007) points out in her study of American Indian tribal police, this became a truism. According to Luna-Firebaugh (2007), "Tribal police administrators and personnel frequently stated that their department personnel saw themselves as the community. These tribal employees saw their job as a relationship with the tribe, an extension of the community. They were responsive to the needs of the community and felt responsible to it. In short, they saw themselves and their department in relationship to each other, not as separate entities, unlike many police in the non-Indian community who tend to view themselves in the role of hired guns whose job it is to fight crime rather than solve problems" (p. 6).

The community policing approach to proactive peacekeeping focuses on the officer's accountability and responsibility to the community. In the adversarial justice approach, police officers are held accountable to the department's chain of command. According to Luna-Firebaugh (2007), "The devolution of power to, and consult with, a broad-based circle of responsible leaders is a common approach to decision making in many Indian communities" (p.57).

A criticism of community policing that is often heard involves fears that a reduction in centralized control can also lead to a lack of appropriate accountability (Luna-Firebaugh, 2007). As Luna-Firebaugh (2007) point out, "Many of the legal challenges to Indian policing are a vestige of an age when the federal government believed that Indian justice was primitive and incomprehensible. For example, U.S. Supreme Court Justice William Rehnquist cited the 1883 *Ex Parte Crow Dog* case when considering tribal criminal jurisdiction over non-Indian residents of a reservation. Justice Rehnquist and a majority of the Court held that such jurisdictional distinction was legally impermissible even though adequate mainstream police services are generally unavailable to Indian Country"(p.127 ).

***"Indians have relationships, Americans have roles"***

*(Luna-Firebaugh, 2007).*

## Law and Culture

One can argue that law is an aspect of culture (Grana and Ollenburger, 1999). According to Becker and Barnes (1961), "An important legal concept for indigenous peoples is that law in the strict sense is found where one group has conquered another and remains in the territory of the conquered as a dominate caste or class. The resulting social stratification is then rationalized. This is also a philosophy of the law relative power argument.

The inferior group is subjected to punishment for any infringement of the interests of their superiors, and thus formal law comes into being" (p. 30).

**Understanding the relationship of the Western criminal justice system is extremely challenging because "political oppression is easier when there is a racial or cultural distinction between the masters and the oppressed. Tyranny will be harsher in a state where all share the same language, culture, and history" (*Sagan, 1995, p. 277*).**

*SOUTHWEST INDIAN BLANKET*

## Conclusion

Are Native American movements towards self-determination positive? The dilemma with this question is the complexity of the Native American culture. According to Nielsen (2009),"Native American 'culture' is not one culture. There are hundreds of North American cultures, all different, all based on unique history, ecology, and values of group" (p. 215). It is accurate, however, to state that many Native American cultures have domestic violence, illiteracy, alcoholism, sexual assaults, suicide, and victimization rates that is much higher than the general population which negatively impact indigenous movements toward self-determination. Self-determination is difficult to achieve when Native Americans have feelings of helplessness and alienation. The alarming social statistics that plague life on reservations are troublesome and it is impossible to find anything positive about high suicide, domestic violence, crime, alcohol use, poverty, and suicide rates as well as the high rates of victimization on Native American reservations. Indian gaming has not solved these social dilemmas on reservations. Much of the research on Native American gaming and crime on reservations is vulnerable to the charge that it is "agenda-driven" because it is sponsored by gaming tribes or those with political interests (Walker, 2001). It remains to be seen what the true impact of gaming on Indian lands will be over the course of time.

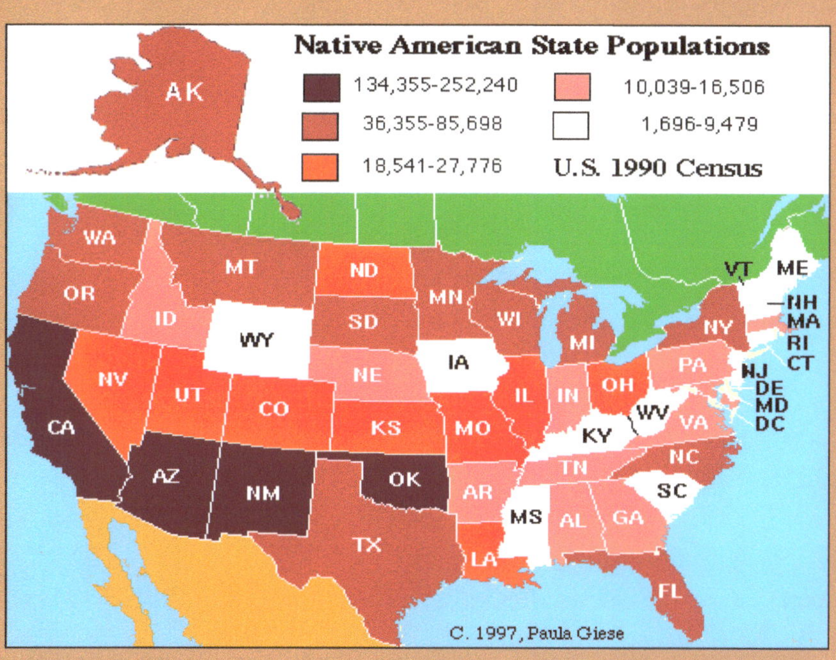


*http://bflibrary.blogspot.com/2011/11/noverber-american-indian month.html*

> ***Native American 'culture' is not one culture. There are hundreds of North American cultures, all different, all based on unique history, ecology, and values of group" (p. 215).—Nielsen 2009***

## References

Alesch, D. 1997. *The Impact of Indian Casino Gambling on Metropolitan Green Bay.* http: www.wpri.org/Reports/Volume10/Vol10n06.pdf. September, 1997.

Amnesty International. 2007. *Maze of injustice: The failure to protect indigenous women from sexual violence in the USA.* New York: Amnesty International.

Archambeault, William. 2009. The Search for the Silver Arrow: Assessing Tribal-Based Healing Traditions and Ceremonies in Indian Country Corrections. *In Criminal Justice in Native American*, ed. Marianne O. Nielsen and Robert A. Silverman. Tucson, AZ: University of Arizona Press.

Bachman, Ronet. 2004. Using Justice Department data to analyze nature and extent of sexual assault against Native American women. Presentation to the U.S. Dept. of Justice Federal- Tribal Working Group on Sexual Assault. Washington, DC, July 12, 2004.

Becker, H., and H. Barnes. 1961. *Social Thought from Lore to Science.* New York: Dover Publications.

Bureau of Justice Statistics. 1997. *Sourcebook of Criminal Justice Statistics*, 1996. NCJ- 165361. Washington DC: Department of Justice.

Diamond, Jared. 1992. The Third Chimpanzee: *The Evolution and Future of the Human Animal.* New York: Harper Collins.

Fairbanks, Kathy. 2003. "Casino San Pablo to be debated Friday in federal courtroom in Sacramento." *PR Newswire*, March 5.

Gould, Larry A. 1999. The impact of working in two worlds and its effect on Navajo police officers. *Journal of Legal Pluralism* 44:53-71.

Gould, Larry, A. 2006. Alcoholism, Colonialism, and Crime. In Native *Americans and the Criminal Justice System.* Boulder, CO. Paradigm Publishers.

Grana, Sheryl, and Jane C. Ollenburger. 1999. *The Social Context of Law.* Upper Saddle River, NJ: Prentice Hall.

Greenfield, Lawrence A., and Steen K. Smith. 1999. *American Indians and Crime.* NCJ 173386. Washington, D.C: U.S. Department of Justice, Office of Justice Programs, Bureau of Justice Statistics.

Grim, Charles. 2002. "Responding through Collaborations." Report delivered at the Tribal Leader Summit on alcohol and Substance Abuse, Albuquerque, New Mexico.

Grinols. Earl L. and David B. Mustard. 2001. "Measuring Industry Externalities: The Curious Case of Casinos and Crime." Working paper, Department of Economics, University of Illinois. Available at http://www.terry.uga.edu/-dmustard.pdf.

Grobsmith, Elizabeth S. 1994. Indians *in Prison: Incarcerated Native Americans in Nebraska.* Lincoln, NE: University of Nebraska Press.

Hamby, Sherry L. 2009. Finding Their Way: Challenges and Resources of American Indian Victims of Sexual Assault. In Criminal Justice in Native America. Tuson, AZ. The University of Arizona Press.

Henderson, Eric, and Scott Russell. 1997. "The Navajo Gaming Referendum: Reservations about Casinos Lead to Popular Rejection of Legalized Gambling." *Human Organization* 56: 294-301.

Huges, Polly Ross. 2003. "Texas Comptroller Takes Gamble for Revenue." Houston Chronicle, April 4, p. 32.

Kahn, M. W. 1982. "Cultural Clash and Psychopathology in Three Aboriginal Cultures." *Academic Psychology Bulletin* 4: 553-561.

Kraus, R. F., and P.A. Buffler. 1979. "Social Stress and the American Native in Alaska: An Analysis of Changing Patterns of Psychiatric Illness and Alcohol Abuse among Alaska Natives." *Culture, Medicine, and Psychiatry* 3: 111-151

Light, S. A. and K. Rand. 2005. Indian Gaming and Tribal Sovereignty: The Casino Compromise Lawrence, KC. University Press of Kansas.

Luna-Firebaugh, Eileen. 2007. *Tribal Policing: Asserting Sovereignty, Seeking Justice.* Tucson, AZ: University of Arizona Press.

Mall, P.D., and S. Johnson. 1993. "Boozing, Sniffing, and Tokin: An Overview of the Past, Present, ad Future of Substance Use by American Indians, " *American Indian and Alaska Native Mental Health Research Journal* 5, no. 2: pp. 1-33.

Mason, W. Dale. 2000. *Indian Gaming.* Norman: University of Oklahoma Press.

McCue, Julia. 2003. "Differing Views." *Maine Sunday Telegram*, March 16, p. 6A

Meyer, Jon'a. 2009. Ha'alchini, haadaah naasdah ("They're Not Going to be Young Forever"):Juvenile Criminal Justice. In *Criminal Justice in Native America*, ed. Marianne O. Nielsen and Robert A. Silverman. Tucson, AZ: University of Arizona Press.

Mihesuah, Devon A. 1993. *Cultivating the rosebuds: The education of women at the Cherokee female seminary*, Urbana: University of Illinois Press.

Mihesuah, Devon A. 1996. *American Indians: Stereotypes and Realities.* Atlanta, GA: Clarity Press.

Muir, Douglas. 2003. "*Casinos in Maine: A Costly Choice.*" Available at http://kittery citizens.org/Gaming_Economics.htm.

National Indian Gaming Commission. 2003. "Tribal Data." Available at http://www.nigc.gov.

Nielsen, Marion and Robert A. Silverman. 2009. *Criminal Justice in Native America.* Tucson, AZ. The University of Arizona Press.

Nielsen, Marianne O., and James W. Zion. 2005. *Navajo Nation Peacemaking: Living Traditional Justice.* Tucson: University of Arizona Press.

Nowlen ,C. 2004. "Casinos Bring Benefits," *Capital Times* (Madison, WI), January 20, 2004.

Peroff, Nicholas. 2006. Indian Gaming and the American Justice System. In *Native Americans and the Criminal Justice System.* Boulder, CO. Paradigm Publishers.

Perry, Steven W. 2002. From Ethnocide to Ethno violence: Layers of Native American Victimization. *Contemporary Justice Review* 5: 231-247.

Perry, Steven W. 2004. *American Indians and Crime: A BJS Statistical Profile, 1992-2002 NCJ 203097.* Washington, DC: I.S. Department of Justice.

Reina, Edward. 2000. "Domestic Violence in Indian Country: A Dilema of Justice." Domestic Violence Report 5, no. 3: 33-48.

Ross, Jeffery Ian, and Larry Gould. 2006. *Native Americans and the Criminal Justice System* Boulder, CO: Paradigm Publishers.

Sagan, E. 1995. *At the Dawn of Tyranny: The Origins of Individualism, Political Oppression and the State.* New York: Vintage.

Saggers, Sherry, and Dennis Gray. 1998. *Dealing with Alcohol: Indigenous Usage in Australia, New Zealand, and Canada.* Cambridge, UK: Cambridge University press.

Schattschneider, E. E., 1960. The *Semisovereign People: A Realist's View of Democracy in America.* New York: Holt, Rinehart and Winston.

Silverman, Robert A. 1996. Patterns of Native American Crime. *In Native Americans, Crime, and Justice*, ed. Marianne O. Nielsen and Robert A. Silverman, 58-74. Boulder, CO: Westview Press

Silverman, Robert A. 2009. Patterns of Native American Crime 1984-2005. In *Criminal Justice in Native America*, ed. Mariane O. Nielsen and Robert A. Silverman, Tucson, AZ: University of Arizona Press.

Sutherly, Ben. 2003. Gaming Complex Dicey, Official Says." *Dayton Daily News*, February 7.

Task Force on Federal, State, and Tribal Jurisdiction. 1976. *Final Report to the American Indian Policy Review Commission.*

Trigger, Bruce G. 1985. *Natives and newcomers.* Kingston: McGill-Queen's University Press.

Tjaden, Patricia, and Nancy Thoennes. 2006. Extent, nature, and consequences of rape victimization: Findings from the National Violence Against Women Survey. Washington, D.C:

U.S. Census Bureau, 2005. American fact finder: S2301: Employment status. http: fact finder.census.gov/servl et/STTable.

U.S. Census Bureau, 2006. *Statistical abstracts of the United States: 2006.* http:// www.census.gov/prod/ www/ statistical-abstract.html.

U. S. Census Bureau. 2007. The *American Community: American Indians and Alaska Natives:* 2004. Washington, D.C: U.S. Department of Commerce.

U.S. Commission of Civil Rights. 2003. *A quiet crisis: Federal funding and unmet needs in Indian Country.* Http://www.usccr.gov/pubs/na0703/na0204.pdf.

Utter, Jack. 2001. *American Indians: Answers to Today's Questions*, 3rd ed. Norman, OK: University of Oklahoma Press.

Walker, Douglas M. 2001. "Kimpt's Paper Epitomizes the Problems in Gambling Research." *Managerial and Decision Economics* 25: 197-200.

Wilkins, David E., and K. Tsiannina Lomawaima. 2001. Uneven ground: American Indian sovereignty and federal law. Norman: University of Oklahoma Press.

Wolf. F. 2001. "Wolf Measure Would Allow State Legislatures to Have Voice in Creation of Gambling Operations on Indian Reservations," Press Release, http://www.house.govwolf/news/2001/06-20Gambling_Indians.html. June 19, 2001.

## ABOUT THE AUTHOR

**Charlie Cernat** is a retired Chief Warrant Officer after 20 years of honorable service in the United States Coast Guard. He received his undergraduate degree in Criminal Justice at Lake Superior State University and his graduate degree at West Chester University. Currently, he is completing his graduate studies in Public Administration. Charlie has a passion for the outdoors and can usually be found hiking or fishing in the northern woods of Michigan.

## PHOTO CREDITS:

Encyclopedia Britannica Kids, Rainforest, Warao House

Examiner: News & Info: These prefabricated native American houses came in many styles, by Richard Thornton

EHow Mom: Native American Lessons for PreSchoolers, Sept., 27, 2011

About Native Americans BLOG, May 27, 2012, About Mandian Houses

Teach Glin.com

Casino News Media, Nov. 19, 2011

The Hubster, Thoughts on Eastie and Beyond, Nov. 29, 2007

Libertarian Republican.net

Think Like A Bride

Navajo Nation Growth, Nov. 13, 2011, VCU Fed ChallengeVirtual Tourist, " Basaic Tuba City"

Virtual Tourist, "Basaic Tuba City Page".

National Center for Education Statistics, Status & trends in the Education of American Indians and Alaska Natives 2008

Banis Blog

Indian Country, Today Media Network.com

Whispering Eagles Trading Post, Native American Lost birds

"Slinging Stones" Blog by D. Barron, Nov. 25, 2011, Tribute to Native Americans-Nations comprised of a Rich Heritage and Culture

Taino-Boricua News

http://www.shantithakur.com/circles.html

MPN Mint Press News, June 11, 2012, Anheuser-Busch, Pabst Brewing Co. Blamed for Alcoholism Damages to Tribes

@ Addiction Blog.org

Buffalo Post, A news blog about Native People and the World We Live In

Walter L. Dutton, Writer, Teacher.

IB. Info Barrel, June 17, 2012, A Brief History of Beer Brewing in the USA

Warrior Nation, Native American Ghost Dance

NMAI, The National Museum of American Indian,

Nov. 11, 2011: A Crow Warrior Vs. The Nazis

Oregon, powered by the Oregonian Native Americans Gather for Pow Wow in Portland to celebrate New Year Without Alcohol, Dec. 31, 2011, by Bill Graves

TH Online.com (powered by Telegraph Herald) Tribes'

latest Casino push hits resistances, July 10, 2012– (Indian Casinos in the US picture).

Zzazzle

Native American Indian Paper Crafts, Squidoo

MMWR Weekly, Aug. 29, 2008, Alcohol—Attributable Deaths and Years of Potential Life Lost Among American Indians and Alaska Natives (CDC)

ATTC—Native Americans & American popular Culture, The Ignoble Savage: The Drunk Injun

NBC News.com

RealHOPE Missions , Native Woman in Reservation

Mendota Mdewakanton

Highland High School Visual Arts

EHow Mom, *Working with Native American Adolescent and Substance Abuse by Saudi Harrison*, eHow contributor

The Fix, Indian Tribe Sues Big Brewers Over Alcohol Problems

Los Angeles Drug Treatment

David McElroy, recovering Political Prostitute: *Do Five Big Beer Companies Force Native Americans to Abuse Alcohol?*

Child Trends Databank

All Power to the People, May 17, 2012

The Buffalo Post

History NYC.com, July 20, 2012, 1904 Native American tandem Breastfeeding photo

Terragalleria.com

Women's Leadership in American History, Motherhood, Hupa (native Americans) mother and Child, CA 1923

MINN POST

Juvenile Justice & Budget Shortfalls, The Crime Report, July 20, 2012

Policymic, Sexual Assault of Native American Women on the Rise

http://bflibrary.blogspot.com/2011/11/noverber-american-indian-month.html

# FORENSIC PORTAL

# NEWS:

- **FALSE CONFESSIONS MAY LEAD TO MORE ERRORS IN EVIDENCE,** *Study Shows*

*ScienceDaily (Nov. 16, 2011)* — A man with a low IQ confesses to a gruesome crime. Confession in hand, the police send his blood to a lab to confirm that his blood type matches the semen found at the scene. It does not. The forensic examiner testifies later that one blood type can change to another with disintegration. This is untrue. The newspaper reports the story, including the time the man says the murder took place. Two witnesses tell the police they saw the woman alive after that. The police send them home, saying they "must have seen a ghost." After 16 years in prison, the falsely convicted man is exonerated by DNA evidence. READ MORE: http://www.sciencedaily.com/releases/2011/11/111116151333.htm

- **New Forensics Research Will Help Identify Remains of Children**

***ScienceDaily(May14, 2010)*** — New research from North Carolina State University is now giving forensic scientists a tool that can be used to help identify the ...READ MORE: http:www.sciencedaily.comreleases/2010/05/100513093733.htm

- **GENES INFLUENCE CRIMINAL BEHAVIOR,** *Research Suggests*

*ScienceDaily (Jan. 26, 2012)* Your genes could be a strong predictor of whether you stray into a life of crime, according to a research paper co-written by UT Dallas criminologist Dr. J.C. Barnes. READ MORE http://www.sciencedaily.com/releases/2012/01/120125151841.htm

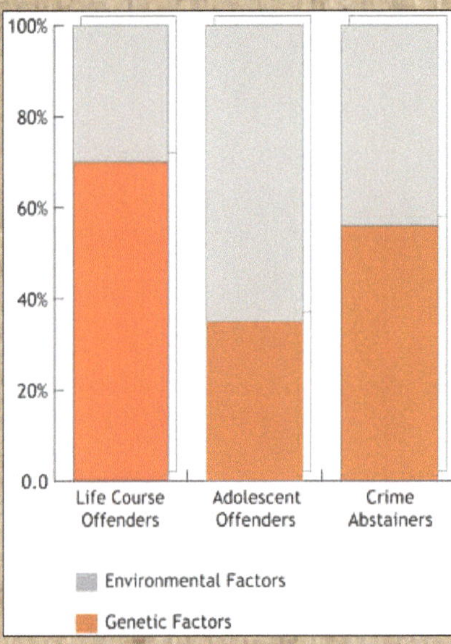


*Genes Show Connection to Crime: UT Dallas criminologist Dr. J.C. Barnes has researched connections between genes and an individual's propensity for crime. Shown is the percentage that genetic factors were found to have influenced whether people became "life course persistent" offenders, "adolescent-limited" offenders, or those who never engaged in deviant behaviors, called "abstainers."*

- **FOR ADOLESCENT CRIME VICTIMS, Genetic Factors Play Lead Role**

*ScienceDaily (May 20, 2009)* — Genes trump environment as the primary reason that some adolescents are more likely than others to be victimized by crime, according to groundbreaking research led by distinguished criminologist Kevin M. Beaver of The Florida State University.

The study is believed to be the first to probe the genetic basis of victimization.
READ MORE: http://www.sciencedaily.com/releases/2009/05/090514153043.htm

- **New Luggage Inspection Methods Identify Liquid Explosives**

http://www.sciencedaily.com/releases/2010/09/100922082339.htm

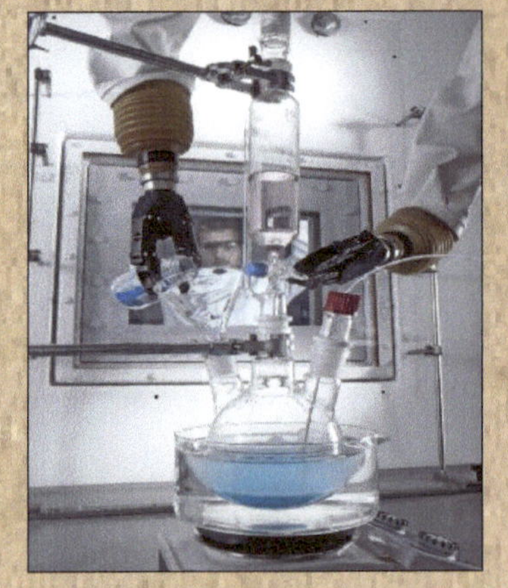

*New luggage inspection methods identify liquid explosives. (Credit: Image courtesy of FraunhoferGesellschaft)*

## ARTICLES:

- **The Corpse as a Scene**

*Mike Byrd Miami-Dade Police Department Crime Scene Investigations*

The corpse at the scene of a brutal homicide can often tell those investigating the death many things. The forensic evidence left behind on the corpse, often times becomes the silent witness against those who commit the most heinous of crimes.

Each investigative entity in a homicide investigation has an assigned duty in the investigation. The crime scene investigator or evidence recovery technician is a support entity in an investigative process. Their assigned task as a forensic specialist deals with the physical evidence.

Their task will include recognizing the items of evidence, documenting any and all evidence encountered, establishing a chain of custody with the items of evidence by isolating and collecting the items, securing the evidence by packaging each individual item in a way that it is not contaminated or lost during the transporting and routing of the items to the laboratory. The items of evidence are distributed to the particular forensic disciplines of the laboratory for analysis.

The success of any investigation starts with a good working relationship between all parties that will be involved. This will require cooperation, commitment, and communications between the detectives, crime scene, medical examiner, and laboratory personnel.
http://www.crime-scene-investigator.net/corpse.html

- **Criminal Responsibility**

  http://www.forensicpsychiatry.ca/crimrespon/overview.htm

- **Forensic Psychiatry**

  http:/www.forensicpsychiatry.ca/paraphilia/overview.htm

- **Lifting Dusty Shoe Impressions from Human Skin:** A Review of Experimental Research from Colorado

  http://www.crime-scene-investigator.net/ShoeImpressionSkin.html

- **Proper Tagging and Labeling of Evidence for Later Identification**

  http://www.crime-scene-investigator.net/tagging.html

## VIDEOS

- **Unreliable or Improper Forensic Science**

  http://www.innocenceproject.org/understand/Unreliable-Limited-Science.php

- **Cell Phone Tracking**

  http://videos.howstuffworks.com/science/forensic-experts-videos-playlist.htm#video-33424

- **Document Forging**

  http://videos.howstuffworks.com/science/forensic-experts-videos-playlist.htm#video-33424

## CASE STUDIES

- **How DNA Contamination can Affect Court Cases**

  http://www.newscientist.com/article/mg21328475.000-how-dna-contamination-can-affect-court-cases.html

- **Forensic Investigation of a Cheating Husband**

  http://www.icsworld.com/Prvate_Investigation_Case_StudiesForensic_Investigation_of_a_Cheaing_Husband_Case_Study.aspx

- **John Wayne Gacey**

  http:forensicsciencecentral.co.uk/johnwaynegacy.shtml

FEATURED ARTICLE :

# JACK THE RIPPER

BY KATHY GAHAGAN AND ROSS NEELY

*The East End of London*

The identity of Jack the Ripper, often regarded as one of the world's most notorious serial killers, has remained a topic of great debate for over a century.

As one of the most popular and controversial serial killers of all time, the unknown murderer who terrorized London's Whitechapel area between August and November 1888 has baffled countless researchers, historians, law enforcement officials, and investigators for over a hundred years. Although many individuals have presented evidence claiming to reveal the true identity of the killer, no single, concrete conclusion exists. Over the course of time, a variety of theories and countless publications have identified numerous individuals as possible suspects, including a doctor, a barber, an artist, and even a member of the royal family. To date, the unsolved murders committed by this unidentified perpetrator are among some of the most gruesome and heinous acts ever documented.

Although this violent criminal was not the first serial killer, his/her crimes did set the standard for the publicity associated with such murders, capturing the attention of the press like never before (Evans & Gainey, 1998). The inability of London's police forces to apprehend and bring the perpetrator to justice caused resentment, unrest and upheaval on behalf of the citizens of Whitechapel and surrounding areas. This unfortunate outcome, coupled with the loss of critical evidence and documented case notes from over a century ago, have resulted in many factors of this case to be lost forever.

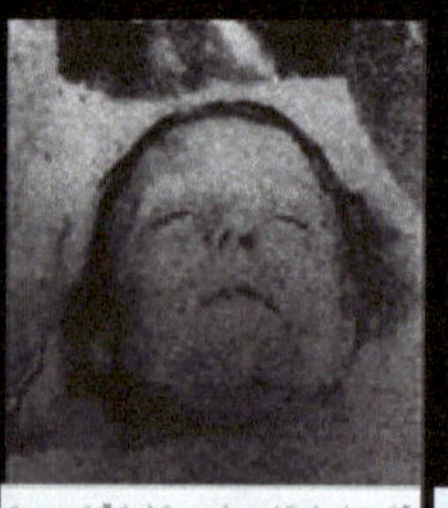
August 31, Mary Ann Nichols, 43

September 8, Annie Chapman, 45

September 30, Lizzy Stride, 44

September 30, Catherine Eddowes, 46

November 9, Mary Jane Kelly, 25

# The Ripper Victims

Between the months of August and November 1888, London's Whitechapel area was shocked by the gruesome murders of five prostitutes which occurred within its city limits. It is important to note that all of the Ripper victims shared similarities that possibly factored into their deaths. In addition to the location of the five murders and the common occupation of the five women, the victims all came from similar socioeconomic standings and lacked personal ties to the community and were estranged from their family and friends.

The murders occurred in the late evening or early hours of the morning in London's crowded East End. Most of the victims were poor, low-level prostitutes known as Unfortunates, all of whom were believed to be intoxicated at the time of their deaths. The women "were on the lowest rung of the social ladder: drunkards whose deaths were little more than a pathetic statistic to all save a few acquaintances and family members"

(Begg& Evans, 2002). In addition, all women had been previously married, widowed, or were not living with their husbands, and had sent their children to live with other people, whether with relatives or an orphanage (Cullen, 1965, p. 5). The Ripper may have specifically targeted these women due to their lack of social and familial ties. Such speculation, certainly not new, suggests that the Ripper was aware, to some greater or lesser degree, with the victims and their histories.

Furthermore, each victim had her throat cut and all except one were severely mutilated after they were killed. "All of the victims were seized apparently from behind and had their throats slit. In all but one case there was some attempt made to mutilate the bodies, though there was no evidence of sexual assault" (Cullen, 1965, p. 3). The victim who was not mutilated was spared that additional fate when the killer was interrupted before carrying out such postmortem acts of disfigurement (Cornwell, 2002). With the exception of this victim, the killer exhibited an escalating trend of violence and mutilation with each successful kill. It can also be noted that prior to the last victim, all women were killed outdoors in public access areas.

## Mary Ann "Polly" Nichols

*Mary Ann Nichols Body Found August 31, 1888*

On Friday, August 31, 1888, Mary Ann Nichols became what is commonly regarded as the first Ripper victim. Her body was discovered in the early morning hours, between 3:40 AM and 3:45 AM (Keppel, Weis, Brown, & Welch, 2005, p. 4). Nichols was a well-known prostitute and alcoholic, which ultimately resulted in the end of her marriage (Douglas & Olshaker, 2000, p. 23). She was last seen working the streets around 2:30 AM in an attempt to pay for her stay at a public lodging house (Holmes & Holmes, 2002, p. 227). Nichols was found "dressed and lying on her back with her clothes pushed up almost to her stomach...Her throat had been slashed from ear to ear" (Keppel et al., 2005, p. 4). "The abdomen had been cut open from the centre of the base of the ribs along the right side and small stabs, apparently with a bladed knife, were made in the lower abdomen" (Wolf, 2008, p.3344).

## *Annie "Dark Annie" Chapman*

*The Ripper's second murder victim **Annie Chapman**. Evidence clustered in yellow oval below her feet.*

Approximately one week after the initial murder of Mary Ann Nichols, Jack the Ripper claimed his second victim, Annie Chapman, on Saturday, September 8th, 1888 (Keppel et al., 2005, p. 6). Annie Chapman was married with two children, but left them, as well as her husband, in order to become a street vendor and a prostitute (Holmes & Holmes, 2002, p. 228). "Dark Annie" was known to Whitechapel as a hostile woman who frequently fought with other women, especially when she drank, which, it was reported, she only did on Saturdays (Holmes & Holmes, 2002, p. 229). At the time of her murder, she was in poor health and suffered from terminal diseases of the lungs and the brain, namely tuberculosis and syphilis (Holmes & Holmes, 2002, p. 229). Chapman was last seen at 11:30 PM when she left a lodging house in search of money in order to afford a place to sleep for the night (Holmes & Holmes, 2002, p. 230).

On September 8th, Chapman's body was found with her dress "pulled up over her head, her stomach ripped open, and her intestines draped over her left shoulder. Her legs were drawn up, knees bent and spread outward. The killer had used a sharp implement, like a surgical knife, to slit her throat…" (Ramsland, 2006, p. 5). The Ripper had placed coins around her body and had removed and taken away her bladder, half of her vagina, and her uterus (Ramsland, 2006, p. 5). "…Two brass rings, evidently wrenched from the middle finger of her left hand, and a few pennies and farthings were laid out neatly at the victim's feet" (Cullen, 1965, p. 51). This murder exceeded the previous murder in its brutality.

*Back Yard No.29 Hanbury St.*

Berner Street

## Elizabeth Stride

Elizabeth Stride, a forty-five year old woman, earned money through prostitution, housekeeping, and sewing (Holmes & Holmes, 2002, p. 231). She was reported to be a foul-mouthed alcoholic who was treated several times for a variety of venereal diseases (Holmes & Holmes, 2002, p. 231). Stride was last seen on September 30, 1888 in the company of a man, who was approximately 5'7" in height, with dark hair, a small mustache, and a long black coat. Furthermore, Stride was overheard by a witness telling the man that she had already earned enough money for that night and was not interested in a sexual encounter (Holmes & Holmes, 2002, p. 233). Stride's body was found at 1:00 AM with a six-inch incision in her neck, but no other mutilations, which caused the police to conjecture that the Ripper had been interrupted during the killing (Wolf, 2008, p. 3344).

## Catherine Eddowes

Mortuary Photo

Later that same night, Catherine Eddowes, a forty-six year old part-time prostitute, was also found murdered in similar fashion (Wolf, 2008, p. 3344). The night before she was murdered, on September 29, 1888, Eddowes was arrested for drunk and disorderly conduct, but was released from jail sometime around 1:00 AM. (Holmes & Holmes, 2002, p. 234). Eddowes was last seen at 1:35 AM by two men who saw her with another man; by 1:45 AM, her lifeless body was discovered (Holmes & Holmes, 2002, p. 236). She was found lying on her back, her abdomen exposed, her throat and right ear cut, the tip of her nose cut off, and her intestines placed over her right shoulder (Holmes & Holmes, 2002, p. 235-236). The police discovered that her body was still warm and that the Ripper took her kidney and uterus from the crime scene (Holmes & Holmes, 2002, p. 236). After the double murder on the night of September 30, Jack the Ripper proved that he was "growing bolder and more expressive of his mental pathology" (Ramsland, 2006, p. 5).

## Mary Jane Kelly

Mary Jane Kelly worked as a maid, a housekeeper, and a nursemaid before becoming a prostitute in a high-class brothel. She was remembered as an intelligent woman with good reading skills (Holmes & Holmes, 2002, p. 237-238). She was twenty-five years old and

*MARY JANE KELLY*

three months pregnant when she was killed by Jack the Ripper on November 9, 1888 (Holmes & Holmes, 2002, p. 239). Kelly was last seen at 3 AM in the company of a man; at 4 AM, someone heard a cry coming from her room, and at 5:45 AM, someone reported seeing a man leave her room. At 10:45 AM, her body was discovered by the manager of the lodging house where she stayed (Holmes & Holmes, 2002, p. 239).

*DORSET STREET*

Unlike the other victims, Kelly was murdered inside of her house and was found "lying naked on her bed with her face hacked beyond recognition. The whole surface of the abdomen and thighs was removed," with some skin placed on a bedside table (Wolf, 2008, p. 3345). "The abdominal cavity was emptied of its organs and the breasts were cut off" and different organs were placed around the body: "the uterus and kidneys were placed under her head…the liver placed between the feet, the intestines by the right side and the spleen by the left side of the body" (Wolf, 2008, p. 3345). This was by far the Ripper's most brutal murder.

# What the murders reveal

*...serial killers may not necessarily be insane, but rather "more cruel than crazy" (Fox & Levin, 2005).*

When attempting to draw conclusions regarding the identity of a serial killer, it is important to first understand what kind of person is being discussed. According to Fox and Levin (2005), serial killers enjoy the thrill, sexual gratification, and dominance they achieve over the lives of their victims. The serial killer rarely uses a gun because this method is too quick and would deprive him/her of his/her greatest pleasure, exalting in the victim's suffering. Their definition also contends that serial killers typically exhibit a sociopathic personality that deprives them of the conscience or guilt needed to guide their behavior. Therefore, serial killers may not necessarily be insane, but rather "more cruel than crazy" (Fox & Levin, 2005).

From the information available, it can be concluded that the perpetrator in the Ripper murders fits this modus operandi. His/her weapon of choice was a knife, or possibly a variety of knives (the exact size and/or type(s) remain unknown). Although the events surrounding the murders produced few, and perhaps less than credible witnesses, their testimony was unsuccessful in aiding the perpetrator's conviction. To date, the identity of the notorious murderer responsible for the brutal killings of Mary Ann "Polly" Nichols, Annie Chapman, Elizabeth Stride, Catherine Eddows, and Mary Jane Kelly remains a mystery.

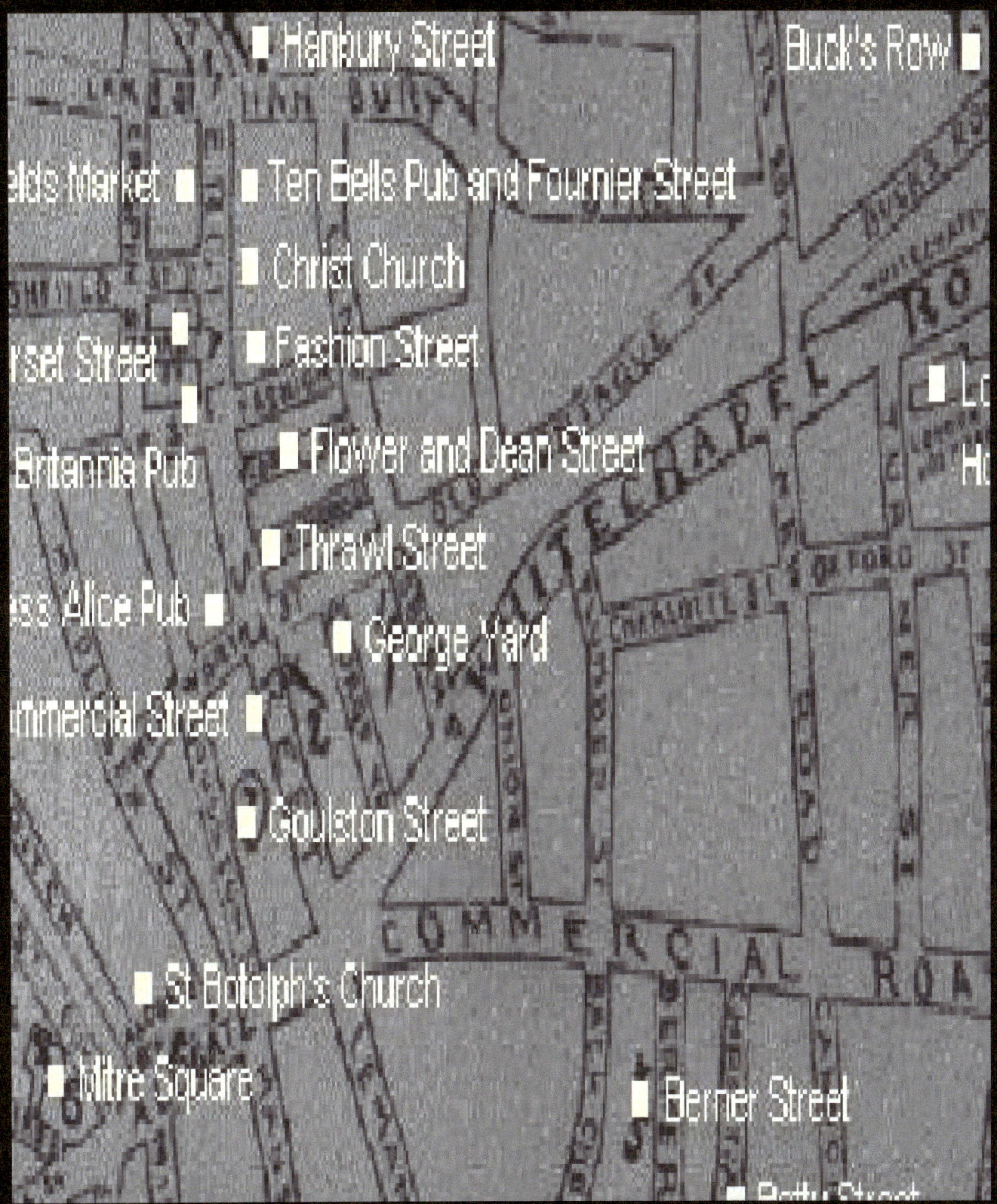


A Map of the East End in 1888 showing the sites of the murders

## Whitecha

There is no doubt that the murders attributed to Jack the Ripper are among some of the most gruesome and heinous criminal acts ever documented. In order to better understand the crimes and familiarize oneself with the perpetrator's possible motives, it is essential to recognize the demographic, social and economic settings of the time.

THE 1800S SIGNIFIED A HISTORICAL PERIOD DURING WHICH ENGLISH SOCIETY WAS EXEMPLIFIED BY TREMENDOUS INEQUALITY AMONG SOCIAL CLASSES. THE CITY OF WHITECHAPEL WAS NO EXCEPTION TO THE MANIFEST DIVISION BETWEEN THE RICH AND POOR. RESIDENTS WERE GEOGRAPHICALLY DIVIDED ACCORDING TO SOCIOECONOMIC STATUS.

Members of the upper class resided in the West End of the city, while members of the lower class were forced to live in the East End. Life for residents of London's East End was extremely difficult. This particular section of the city was considered a breeding ground for alcoholism, prostitution, and crime. It was a small and greatly overpopulated area, a place where poverty ran rampant.

From this information, it becomes easily apparent that crime and poverty

## ndon 1800s

affected many people, especially the residents of the East End. In 1861, "800 thieves, vagabonds, beggars and prostitutes" were living in the East End. By 1889, the area was described as "very poor, lowest class... vicious, semi-criminal" (Cullen, 1965, p. 21). The people in the East End had to endure great oppression. "Close to a million people lived crammed together in tenement houses...Work was scarce, and women frequently turned to prostitution to feed their families" (Goldman, 2001, p. 24). The living conditions of Whitechapel forced people to live in squalor and degradation.

Moreover, out of the 900,000 people living in the East End of London, the 90,000 people who resided in Whitechapel had the highest percentage of poverty (Rubinstein, 2000). It was home to poor immigrant families, most of whom resided in dilapidated housing quarters with as many as eight people per room. "Those lucky enough to have a place to live were crammed into dirty and primitive accommodations without even the semblance of privacy" (Douglas &Olshaker, 2000, p. 22). On the other hand, countless others were forced to live "a day-to-day existence- on the streets, in grim and notorious public workhouses, or in the...filthy "doss-houses," which offered a bed for around fourpence a night, paid in advance" (Douglas &Olshaker, 2000, p. 22).

*Whitechapel Market in the1800's*

*The conditions of Whitechapel were so awful that women were not even looked down upon for turning to prostitution to support themselves and their families.*

The lack of available employment in the East End understandably could not reduce the area's poverty. "...In Whitechapel, skilled jobs were scarce and disease was rampant" (Douglas &Olshaker, 2000, p. 23). Those who were fortunate enough to find employment worked for meager earnings in deadly factories which emitted a constant flow of toxic gases into the air, making visibility nearly impossible(Curtis, 2001). In addition, a large number of the men in Whitechapel were reduced to working "on a casual day-to-day basis;" while women who worked often acquired jobs that entailed cleaning, doing laundry, or needlework, and the women frequently turned to prostitution to make more money (Wolf, 2008, p. 3343). The Metropolitan Police found that there were approximately 1200 prostitutes who lived in Whitechapel in 1888, the year of the Jack the Ripper murders (Wolf, 2008, p. 3343). The conditions of Whitechapel were so awful that women were not even looked down upon for turning to prostitution to support themselves and their families. "Overcrowding, unemployment, disease, and other social maladies affected thousands of people, particularly women...Because conditions were so bad, no one thought ill of the women who had to resort to the streets for subsistence" (Holmes & Holmes, 2002, p. 224).

Despite the unforgiving conditions, one problem that the residents of Whitechapel did not have to concern themselves with was murder, until the emergence of Jack the Ripper and his horrific killing spree. "...Out of the sixty-eight murders committed in London in 1886, and of the eighty committed in 1887, not a single one was committed in Whitechapel" (Schmid, 2005, p. 33). The infrequency of previous homicides in the East End, coupled with the grotesque nature of the killings, made the Ripper forever infamous (Rubinstein, 2000). The murders caused such a sense of widespread fear and panic amongst the public, many residents of Whitechapel did not want to go out at night in fear for their safety. According to the landlord of The Star and Garter, "People aren't going out at night anymore...Since the killings, I hardly get a soul in here of a night"" (Cullen, 1965, p. 2). For the residents of Whitechapel, these terrible murders caused them to endure even more suffering and oppression in their daily lives.

The extreme overpopulation of the urban district, in combination with the horrific health conditions caused by poor drainage and inadequate sanitation, created an environment in which disease claimed countless lives and death and starvation were common occurrences. The extreme poverty, high alcoholism rates, and lack of parental supervision in the East End turned the streets into chaotic stomping grounds for criminals, prostitutes and beggars (Curtis, 2001). The overall sense of darkness and depravity within the East End epitomized the repugnant character of Jack the Ripper. It can also be noted that during this same time, and only a few short miles away, those residing in the West End experienced a lifestyle quite contrary to those in the East End. Residents of the West End lived their lives filled with great wealth and prosperity. It was not uncommon for these residents to "slum" or take riding tours through the impoverished East End, nor was it uncommon for wealthy men to frequent brothels in this area (Koven, 2004).

*The overall sense of darkness and depravity within the East End epitomized the repugnant character of Jack the Ripper.*

# The Ripper Suspects

The gruesome details surrounding the murders make it overtly apparent that the acts were ones of great violence and cold blooded murder. The ever elusive perpetrator succeeded in bringing an overall sense of panic and fear to the citizens of London and others around the world. The nature of these murders, along with their subsequent media attention, produced a variety of theories surrounding the Ripper's identity. The killer's true identity, however, remains a mystery.

The inability of London's police forces to effectively apprehend the perpetrator can be attributed to a variety of factors. For example, crime scene photography, an established evidence chain of command, and proper maintenance of official records were not a common practice during this time. Poorly trained officers, crime scene investigators, medical examiners, and an overall lack of forensic science also contributed to the absence of reliable evidence and records. In fact, forensic science laboratories were not yet developed, so forensic examinations were only conducted on the bodies and letters which the killer allegedly sent to the police and media (Eckert, 1981). Such shortcomings can be regarded as twofold. Not only did they produce a lack of viable evidence at the time, but they also directly impacted more recent efforts to solve the mystery. These aspects do not include the fact that, in order for such material to be readily available today, it would have had to survive not one, but two World Wars, a series of police precinct moves, multiple changes in power, and over a century's worth of decomposition.

For the purposes of this article, the list of possible suspects has been narrowed down to those who have been the most 'popular' over time. The suspect list includes those individuals which these researchers believe were the most likely to have been involved in the murders. It is important to note that the following is not a comprehensive list of all possible suspects. Additional suspects exist, all of whom are not mentioned in the following.

## *Prince Edward Albert*

Prince Edward Albert, commonly known as Prince Eddy, was the Duke of Clarence and the grandson of Queen Victoria, as well as a suspect in the Jack the Ripper murders (Holmes & Holmes, 2002, p. 241). During his youth, he was tutored at Cambridge where it was reported he had difficulty concentrating on material due to diminished capacity and was known to frequent homosexual establishments. Although Prince Eddy did eventually marry, his early death prevented him from ever being crowned king. Research indicates that he was a man of inadequate education, low intelligence and poor motivation. Due to a hereditary condition, he was also partially deaf. The premise of the Ripper theory contends that Prince Albert "suffered from effects of syphilis on the brain as a result of his debauching and that he used to slum in Whitechapel and pick up lowly women. The dementia caused him to kill some of these women for sport..." (Douglas & Olshaker, 2000, p. 67-68). Other supporting evidence suggests that because he was a deer hunter, he had the necessary skills to disembowel the victims (Douglas & Olshaker, 2000, p. 68).

Yet another theory behind Prince Albert as a suspect is that he secretly married and had a baby girl with Annie Elizabeth Crook, a poor, lower-class woman who was also Catholic, which by law forbade Prince Albert to marry her because she was not of the Church of England (Douglas & Olshaker, 2000, p. 68). To prevent this scandal from becoming public, the royal family sent Crook to an asylum; however, the baby's nursemaid was Mary Jane Kelly, who told her friends, Polly Nichols, Annie Chapman, Elizabeth Stride, and Catherine Eddowes, of the scandal and tried to blackmail the government (Douglas & Olshaker, 2000, p. 68). The royal physician, Sir William Gull, was then sent to find the women and kill them (Douglas & Olshaker, 2000, p. 68)

In the two previously mentioned theories, there are many flaws. For instance, the prince had alibis for each of the murders through eyewitness accounts, royal diaries, and court circulars (Douglas & Olshaker, 2000, p. 68). A second flaw in the theories is that a person trained to appropriately interact with the public, such as the prince, may not have been able to commit the heinous Ripper crimes and continue to properly function and interact with people. Instead, the Ripper would have had to be someone who did not know "how to interact with women, and whatever his personal hang-ups or character flaws, Prince Albert Edward would have been trained to this social grace" (Douglas &Olshaker, 2000, p. 68-69). Furthermore, the killer seemed to be disorganized and paranoid, and if the prince was indeed the killer, it is not likely that he would have risked traveling to a foreign neighborhood where he could have been "recognized with the intended purpose of mutilating women he'd never met" (Douglas & Olshaker, 2000, p. 69).

Authors such as Phillipe Jullien and Thomas Stowell have contemplated the prince to be the Ripper murderer, despite the police not considering him a suspect at the time of the crimes. In a 1970 publication of the *Criminologist*, Dr. Thomas Stowell claims that the royal family knew Eddy was responsible for the Whitechapel murders, but made no attempts to restrain him (Stowell, 1970). Many aspects of his life are simply not known, and there are a multitude of theories surrounding his death. It is well documented that Prince Eddy died in January of 1892 at the age of twenty eight. His death was attributed to the pneumonia and flu epidemic which swept the country during this time. However, others such as Frank Spiering (1978) suggest that Lord Salisbury and Albert Edward had Eddy killed by a morphine overdose. Stowell claims he died from softening of the brain due to syphilis, while another theory suggests he did not die until 1930 when he was locked away in the Osborne House, a compound for the insane (Morley, 2005).

Regardless of the manner or exact date of his death, historical court circulars indicate that at the time of Mary Ann Nichols' murder, Prince Eddy was at Dandy Lodge Grossmont in Yorkshire. When Annie Chapman was murdered he was in York, and the night Elizabeth Stride and Catherine Eddowes were murdered, he was in Scotland. Moreover, when Mary Kelly was killed, the prince was at Sandringham (Morley, 2005).

Based on these facts it is the conclusion of these writers that unless all the official records were forged, Prince Eddy was not responsible for the murders committed by Jack the Ripper. He may have had the royal family on his side, but to successfully forge official records would constitute treason at the highest level and he would have had to have multiple co-conspirators, something that seems unlikely.

## *Charles Lutwidge Dodgson (Lewis Carroll)*

Another Ripper suspect Charles Lutwidge Dodgson, more commonly known by his pen-name, Lewis Carroll, was born January 27, 1832. For our purposes, Dodgson will be referred to as Lewis Carroll. Carroll was described as being a handsome man, around six foot tall with wavy brown hair, blue eyes, and a slender build. In 1849, Carroll traveled to Oxford and enrolled in ChristChurch (his father's old college). He was socially competent, intelligent, persuasive, manipulative and attractive to women. He was deeply involved in academics, enjoyed the theatre and the arts, and became a distinguished photographer with a passion for the nude. Carroll also enjoyed a variety of open and intimate friendships with a number of different women (Morley, 2005).

In 1861, he became a deacon of the Anglican Church, but he refused to become a priest. Instead, Carroll became a well-known author and continued to teach at ChristChurch until 1881. Through a series of publications, the most popular of which was *Alice's Adventures in Wonderland* (1865), he experienced a substantial amount of wealth and fame which enabled him to live in comfort until his sudden death from pneumonia in January of 1898 (Leach, 1999).

Some researchers have stated that Carroll was responsible for the Whitechapel murders of 1888. Perhaps most notably, author Richard Wallace recently claimed to have decoded several anagrams in the works of Carroll, linking them to the letters written by Jack the Ripper. Wallace argues that although Carroll was a grown man, he had the mind of a child and used his writings to document the actions of his secret life. For example, he converts Carroll's passage:

> "So she wondered away, through the wood, carrying the ugly little thing with her. And a great job it was to keep hold of it, it wriggled about so. But at last she found out that the proper way was to keep tight hold of itself foot and its right ear."

> And translates it into: "She wriggled about so! But at last Dodgson and Bayne found a way to keep hold of the fat little whore. I got a tight hold of her and slit her throat, left ear to right. It was tough, wet, disgusting, too. So weary of it, they threw up - Jack the Ripper" (Wallace, 1996).

Wallace also suggests that Lewis was a closet homosexual who hid his true sexual desires and exerted his distaste for women through the murdering of prostitutes.

Throughout researching the life of Lewis Carroll, as well as the theories which implicate his involvement in the Whitechapel murders, these writers found Wallace's work to be the most applicable. That being said, it would be both ill-informed and inaccurate of us to conclude with reasonable sensibility, that Carroll was responsible for the Ripper murders. Although one can credit Wallace for his dedication to his work, to draw such a conclusion based on the interpretations of fictional writings would be unfounded. There is simply no concrete evidence presented by Wallace that successfully links Carroll's fictional writings (perhaps entirely meaningless) to the journal of a killer. Therefore, these writers find Carroll's implication in the Ripper murders to be inaccurately based on unsubstantiated evidence. For one to reasonably conclude that he was responsible for some of the most heinous crimes ever committed, based largely on such farfetched ideas and biased interpretations, the writers find quite disturbing.

## *Montague Druitt*

Montague Druitt was quite a popular suspect in the Jack the Ripper murders; in fact, for over twenty years, he was most likely the number one suspect (Rubinstein, 2000). Druitt was a schoolteacher and a barrister, but, as a teacher, he was believed "to have gotten into trouble for sexual advances to some of the students (Douglas & Olshaker, 2000, p. 78). After being dismissed as a teacher, Druitt committed suicide by drowning in the Thames River around November 30, 1888, three weeks after the murder of Mary Jane Kelly (Rubinstein, 2000).

Druitt's status as a Ripper suspect can be credited solely to Sir Melville Macnaghten, "the onetime head of the Criminal Investigation Division at Scotland Yard" (Cullen, 1965, p. 4). Macnaghten thought Druitt was a doctor, about forty-one years of age, whose own family believed him to be Jack the Ripper (Cullen, 1965, p. 249, 251). However, Macnaghten's belief proved to be wrong. Druitt was not a doctor, but a barrister, and he was thirty-one years of age at the time of his death.

"Extensive research has failed to find evidence clearly linking him to the Ripper crimes"(Rubinstein, 2000). "Aside from his untimely but convenient death, nothing really ties him to the crimes, including any known association to Whitechapel. There is no evidence of violence in his background, and a man doesn't just jump full-blown into the kinds of crimes" that Jack the Ripper committed (Douglas & Olshaker, 2000, p. 78). Therefore, Druitt has been disproven to be Jack the Ripper. The evidence against him, particularly that of Macnaghten, proved to be incorrect and of little theoretical basis.

## Aaron Kosminski

Also named as a Jack the Ripper suspect by Macnaghten was Aaron Kosminski. In 1894, Macnaghten claimed Kosminski to be a viable suspect, stating that he was "a Polish Jew" and a Whitechapel resident who became insane, had a great hatred of women, especially prostitutes, "had strong homicidal tendencies," and was sent to an asylum in 1889 (Begg& Evans, 2002). Adding to Macnaghten's belief that Kosminski was a viable suspect was a witness who claimed that he was the Ripper. However, this witness, after learning that Kosminski was also Jewish, refused to testify against him in court "for fear of hanging a fellow Jew" (Douglas & Olshaker, 2000, p. 76; Rubinstein, 2000).

The evidence against Kosminski does not prove that he was the Ripper. Even though Kosminski "carried a knife, seemed threatening, disliked women, and was odd, strange, or outright insane," he still could not have been Jack the Ripper (Begg & Evans, 2002). There is no evidence that Kosminski was ever dangerous or violent, other than what was claimed by Macnaghten (Rubinstein, 2000). Furthermore, while in the asylum, Kosminski "was often dissociative but not violent and never gave any indication of being the Ripper (Douglas & Olshaker, 2000, p. 79). "...His condition when committed doesn't suggest that he was someone capable of killing and mutilating women or that he was the sort of person even the most desperate prostitute would have gone off with" (Begg & Evans, 2002). Therefore, it does not seem possible that Kosminski could have been Jack the Ripper.

## Dr. Thomas Neill Cream

Dr. Cream has been a popular suspect for Jack the Ripper because of three words he said before being hung to death. These words were: "I am Jack" (Foran, 2006). Cream was born May 27, 1850 in Scotland. After graduating from McGill College in Montreal, Canada, he began practicing medicine. When he discovered his lover was pregnant, he attempted to perform an abortion on her; however, after contracting bronchitis she died "of what appeared to be consumption," causing her doctor to contemplate whether her death was connected to medicine that Cream sent her (Foran, 2006). Cream soon fled England and set sail for the U.S. He began conducting illegal abortions and was responsible for the deaths of three more women. One died in Ontario from chloroform, another in Chicago as a result of a botched abortion, and a third woman who died from taking medicine prescribed by Cream (Foran, 2006). He was arrested for the crimes, but later released due to insufficient evidence. It was also reported that he enjoyed frequenting the local prostitutes in the Chicago area, contracting syphilis as a result.

Cream then began having an affair with a married woman and was charged with the murder of her husband, Daniel Stott, after his body was discovered on June 14, 1881 (Morley, 2005). He was subsequently found guilty of second-degree murder for the murder of sixty-one year old Stott (Foran, 2006). Cream was sentenced to life in prison and was sent to the Illinois State Penitentiary. During his time in prison, it is believed his illness further deteriorated his mental state. He was often heard ranting about seeking revenge against women should he ever be released (Rosenhek, 2005). Apparently his intentions were sincere, and unfortunately for several women, he was released in 1891 after serving ten years of his sentence. Cream soon returned to London where, in 1892, he murdered four prostitutes through strychnine poisoning (Douglas & Olshaker, 2000, p. 71).

After a failed attempt to blackmail and extort a neighbor by identifying him as the murderer, Cream was arrested in June Cream supposedly boasted to fellow inmates about numerous other murders for which he had not been found guilty. As he was being led to the gallows, he uttered his famous last words, "I am Jack...."

....However, the floors dropped and he was hanged at Newgate Prison (November 1892) before finishing the statement (Morley, 2005).

It is important to note that Cream was released from prison in 1891, three years after the Ripper murders took place, which largely dismisses him as Jack the Ripper (Foran, 2006). However, some have suggested that Cream successfully bribed his way out of prison, traveled to London to commit the murders, then returned to Illinois to recommit himself to prison (Bell, 1974). Although not entirely impossible, this theory seems highly unlikely. If Cream had managed to bribe his way out of prison in the U.S., and make his way to London, why then would he ever return? What benefit would this have? Perhaps he did manage to create the perfect alibi for the murders. Based on the research conducted by these writers, it was concluded that Cream did not pull off such an elaborate scheme and remained incarcerated during the time of the Ripper murders. In further support of this statement, records from Joliet Prison indicate that Cream (prisoner no.4374), was imprisoned from November 1st, 1881 through July 31st, 1891 (Morley, 2005). Cream was indeed a murderer, but was not responsible for the murders committed in Whitechapel by the Ripper. All of his victims died from poisoning, or as a result of botched abortion attempts. Furthermore, none of his victims were brutally murdered or mutilated in the manner which was the Ripper's modus operandi.

## George Chapman

George Chapman, born Severin Klosowski, found work as a hairdresser and barber in Whitechapel upon his arrival in London (circa 1887). His activities over the next few years were anything but admirable. He would marry multiple times, with the end of a marriage often meaning the death of a wife. In 1981, Chapman briefly moved to the U.S. only to return to London in 1982. Although he was not initially-charged in connection with their murders, his next two wives died of apparent health complications. When his final wife also turned up dead, an investigation into her death was conducted. Her death was then ruled a murder from ingested poison, and Chapman was convicted and hung in April of 1903 (Morley, 2005).

His conviction caused many to consider Chapman a likely suspect in the series of unsolved Ripper murders. In fact, many officials in Scotland Yard focused on Chapman as a primary suspect during the time of the murders and several arguments have been made in support of Chapman as the killer. After all, not only did Chapman have a long history of murdering women, but he was also present in London during the time of the murders. The first argument in support of this theory is that Chapman arrived in Whitechapel shortly before the first Ripper murder and that he was in close proximity to all of the murders (Cullen, 1965, p. 230). The second argument states that the description of the man seen with Mary Kelly before her murder was similar to that of Chapman (Cullen, 1965, p. 230). The final argument identifying Chapman as the Ripper is that the Ripper letters contained "Americanisms," which Chapman would have learned while he was in America for two years and which would have helped Chapman pose as an American (Cullen, 1965, p. 230).

However, some of the aforementioned arguments can be quickly refuted. For example, Chapman did not visit the United States until 1890; therefore, he could not have been proficient in the use of "Americanisms" at the time of the murders (Cullen, 1965, p. 231). Additionally, the descriptions provided to the police of the men seen with the victims before their deaths were all of middle-aged men; Chapman was only twenty-three years old at the time of the murders and could not have looked to be middle-aged (Cullen, 1965, p. 231). Also, Douglas and Olshaker (2000) claim that "there is no way a man hacks apart five...women...then resumeshis homicidal career as a poisoner."

One cannot deny that Chapman was a man of great suspicion and immoral character, but to say with certainty that he was the Ripper murderer is another claim entirely. When considering the possibility, two main inconsistencies were presented to these researchers. The first questionable correlation lies within the manner in which Chapman murdered his victims. It is known that he slowly poisoned his victims over a lengthy period of time, and did not exert any kind of physical violence towards them. Chapman neither engaged in any documented acts of bloody murder, nor did he engage in any acts of postmortem mutilation.

Although most serial murders operate using a similar modus operandi during the course of their killings, it is not unheard of for one to change such tactics over time. For example, a more recent American serial killer and contract hit man named Richard Kuklinski began by murdering and sometimes mutilating his victims. He continually learned new tactics through association with other criminals and incorporated the use of cyanide into his killing strategy, as an example. When Kuklinski murdered with cyanide, there was no violence and no mutilation, only death. Therefore the question remains: was it possible for Chapman to murder his wives with no acts of physical violence while also murdering strangers in a completely separate and brutal fashion? It would appear that such a possibility is not out of the question; however, it does seem a rather unlikely phenomenon.

The other factor which seems to be unexplained is the time between the murders of Chapman's wives and the Ripper murders. If Chapman was indeed responsible for the series of Ripper murders in 1888, why then did he suddenly stop murdering only to strike again (in an entirely different fashion) in 1897? Would Chapman have been able resist the urge, all the while remaining in London after the initial cold blooded murders in 1888?

The FBI's Behavioral Science Unit's (BSU) differentiation between spree and serial murders may provide some assistance in answering such questions. Their classification is based on whether or not the offender "cools off" between crimes. The BSU classifies a spree killer as an individual who commits his/her acts over a period of time, usually several days, wherein his/her activities include the planning and execution of the crimes and evasion of police. In contrast, the serial killer may continue to kill over a period of months or years and often

has long lapses between murders. During this time, he/she maintains a seemingly normal life, often spending time with family and/or working (Fox & Levin, 2005). According to this theory, it may in fact have been possible for Chapman to achieve such an existence. Perhaps the answer to such questions will never be entirely known.

Suppose for a moment that Chapman did in fact manage to wait nine years between murders, and was responsible for both series of murders. Such actions still fail to explain the two entirely contradictory styles used to commit the murders. Therefore, it is the conclusion of these writers that Chapman was not the Ripper murderer.

Based on the knowledge and psyche of serial killers, it can be concluded that Chapman would not have been able to revert to less drastic killing strategies after experiencing the "thrill" involved in the up close and personal Ripper murders. A variety of publications, focused on the development and psychopathology of serial killers, support the position that deviant behavior feeds upon itself. In other words, what was once viewed as satisfying and thrilling to a perpetrator will eventually cease to satisfy him/her, causing him/her to graduate to more violent and extreme practices (Turvey, 1999; Giannangelo, 1996). This information withstanding, it is extremely unlikely that Chapman began with violence and mutilation, only to regress to kill again by slowly poisoning his victims. Furthermore, if Chapman was a man who could not even get away with poisoning his victims during a time when forensic science was a phenomenon largely read about in Sherlock Holmes novels, and medical examiners were rudimentary in their professional skills, he certainly was not capable of evading capture while executing the Ripper murders.

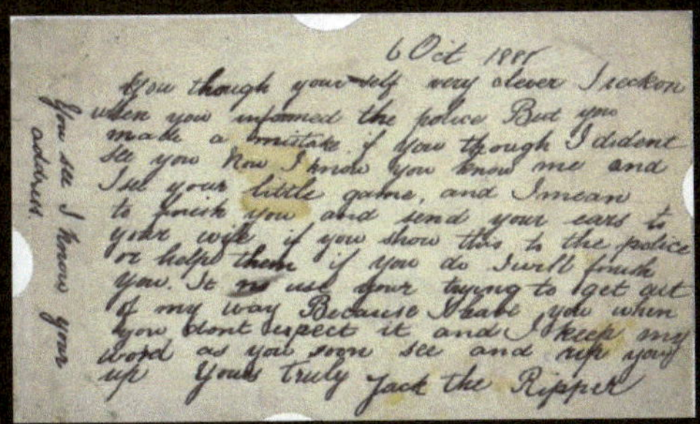

6 Oct 1888

You though your-self very clever I reckon
when you informed the police But you
made a mistake if you though I didnt
see you Now I know you know me and
I see your little game, and I mean
to finish you and send your ears to
your wife if you show this to the police
or help them if you do I will finish
you. It no use your trying to get out
of my way Because I have you when
you dont expect it and I keep my
word as you soon see and rip you
up Yours truly Jack the Ripper

You see I know your address

## Walter Sickert

Another popular suspect indicated in the Ripper murders was the famous painter, Walter Sickert. In her book, *Portrait of a Killer: Jack the Ripper- Case Closed*, Patricia Cornwell (2002), claims to have cracked the previously unsolved mystery by identifying Sickert as the Ripper murderer. She asserts her conclusion through the use of criminal profiling, detective investigation and evidential findings. It should be noted that prior to the publication of this work, Cornwell authored a number of fictional crime novels but did not publish works of actual fact. There is no doubt that she presents a variety of interesting and compelling arguments, and her extensive research and dedication to this piece cannot be denied. On the other hand, the actual amount of credibility and extent to which her proposed theory holds true is debatable.

Cornwell's theory was supported by her belief that Sickert's sketches, letters, and DNA all proved that he was the Ripper. Through examination of the paintings, and analysis of the handwriting and DNA on the Ripper letters and on the letters written by Sickert, Cornwell claimed that Walter Sickert was Jack the Ripper (Cornwell, 2002). However, Cornwell's theory has many fallacies and her reasoning behind her claims of having identified Jack the Ripper as Walter Sickert appear to be not only incorrect, but they also seem to be a product of poor investigative work.

A considerable amount of Cornwell's thesis focuses on her conclusion that Sickert was responsible for most of the taunting Ripper letters sent to the police and media during the time of the murders. With the help of a number of forensic scientists, Cornwell claims to have matched a sequence of mitochondrial DNA found on several Ripper letters to that on letters written by Sickert. While Cornwell admitted that she could not be certain that the tested materials contained DNA from Sickert and the Ripper, she still maintained the assumption that the results obtained from the analysis, mainly that a DNA sequence from a Ripper letter was also present in letters from Sickert, proved that Sickert was responsible for the five Ripper murders. However, Cornwell did admit that the analyzed DNA was the oldest ever tested and that, as a result, it could not definitively prove that the DNA was that of both the Ripper's and Sickert's (Cornwell, 2002, p. 11-12).

When considering these findings, several important factors should be addressed. First of all, Cornwell's attempts to match nuclear DNA sequences (the standard for all DNA testing) were unsuccessful. This type of DNA, sequences of DNA because it is passed down from both parents. In contrast, mitochondrial DNA is found outside the nucleus of the cell, therefore limiting the results because it is passed down only from the mother.

Aside from the fact that mitochondrial DNA cannot be used to positively identify a single individual, attempting to test DNA material over a century old has rarely, if ever, been done before. Although Cornwell uses this evidence as a foundation for her thesis, these writers are hesitant to arrive at such a conclusion. Upon examination of her statements, we find it essential to note several potential errors and shortcomings. For example, Sickert's exact DNA profile is not known, the DNA used could have been contaminated, and Cornwell's research lacked standards for comparison. Such potential limitations can be observed within her own writing. For example, Cornwell writes, "the mixture, poor quality of samples, and lack of references from the individuals in question weaken these data and mark them as 'questionable'" (Cornwell, 2002).

Furthermore, as stated by Terry Melton, an expert in the analysis of mitochondrial DNA, the letters that were analyzed would have also included DNA from all of the people who have touched the letters since the time in which they were written, over 115 years ago (Murray, 2004, p. 55). Cornwell disregarded this fact in her book and expressed her findings as if they proved Sickert was Jack the Ripper, despite the fact that another letter had DNA from a different Ripper suspect, Montague Druitt. She quickly ruled out Druitt as a suspect because, in her opinion, no evidence has linked him to the murders and the murders continued after his suicide in in 1888 (Cornwell, 2002, p. 12). Cornwell did not convincingly prove this part of her case, from the perspective of the writers, especially because she failed to mention the limitations of the DNA analysis.

# Letters from Jack the Ripper

*"One of the first problems with this conclusion lies within the fact that most, if not all, Ripper letters written to the police and the media are believed to be hoaxes".*

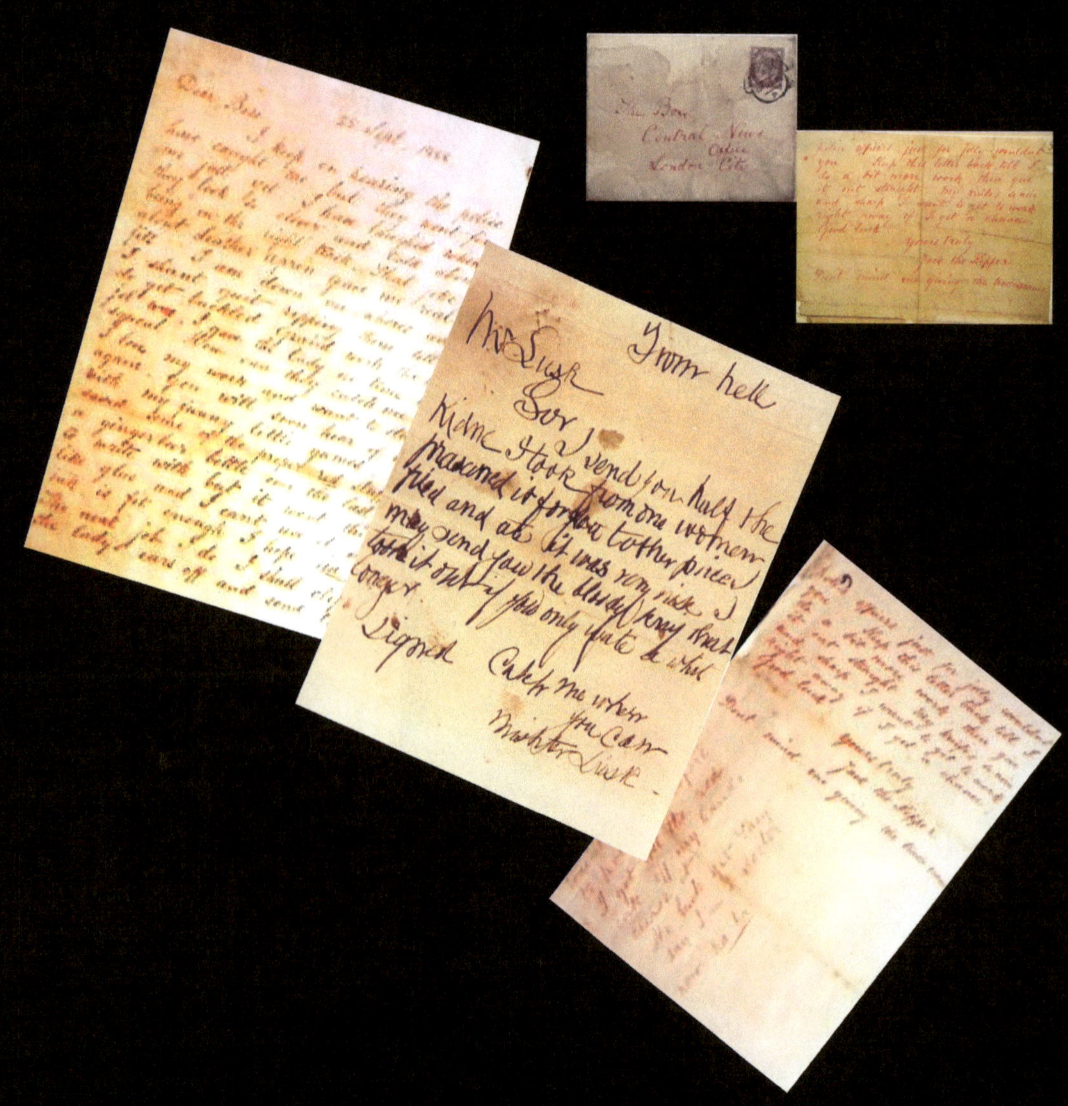

Besides claiming to have linked Walter Sickert to the Ripper letters through matching mitochondrial DNA sequences, Cornwell contends that he possessed the capability of disguising his handwriting in a variety of ways. According to Cornwell, he wrote far more letters than most are inclined to believe, "because one makes a mistake to judge Walter Sickert by the usual handwriting comparison standards" (Cornwell, 2002).

During the time of the murders and the years which followed, well over three hundred letters were written by anonymous authors claiming to be the Ripper. Not only were these letters sent from locations all across England and countries overseas, but also varied in handwriting styles, spelling and grammar. Some letters were written in clear, precise and educated fashion while others contained misspellings of the most basic words.

Cornwell states that through his baffling varieties of papers, pens, postmarks, paints, disguised handwritings, and constant moving around, Sickert created an investigative chaos.

Another apparent fallacy within Cornwell's work lies within her claim to have successfully matched specific watermarks in both Sickert's letters and those sent to the police and media. Upon review of hundreds of letters, she discovered that watermarks from the A. Pirie & Sons, Joynston Superfine and Monckton's Superfine were present in both types of letters. Cornwell then surmised that Sickert was responsible for sending the letters and presents her findings as overwhelming evidence linking him to the murders. Before jumping to such a conclusion, this supposed phenomenon can be addressed as a simple matter of probability.

**...it seems unreasonable to believe that a single author was responsible for writing even a portion of the letters. In fact, forensic documentation examiners in London recently concluded that journalists working within the Central News Agency were responsible (most likely) for writing the first Ripper letter. The original "Dear Boss" letter then opened the floodgates for hundreds of similar hoaxes to follow, and mentioned the name "Jack the Ripper" for the first time (Eyres, 2000).**

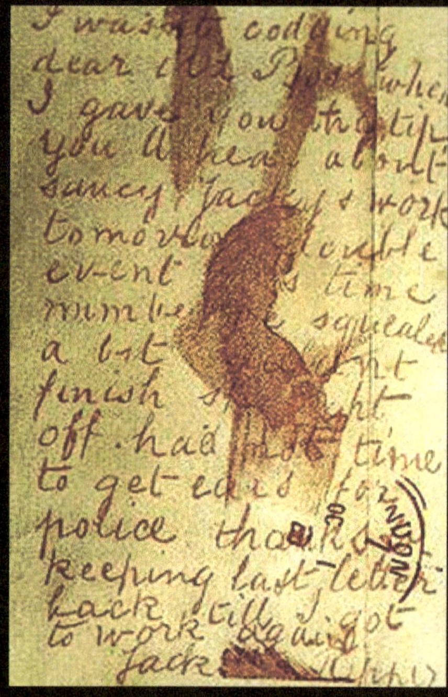

I was[illegible] codding
dear [illegible] when
I gave you the tip
you'll hear about
saucy Jacky's work
tomorrow double
event this time
number one squealed
a bit [illegible]
finish straight
off. had not time
to get ears for
police thanks for
keeping last letter
back till I got
to work again
Jack the Ripper

During the time the Ripper letters were sent, only around ninety paper mills were in operation in England. Of these mills, A. Pirie & Sons was the largest, while the Joynston and Monckton's Superfine's were two significant others (Cohen, 2002). Keeping this information in mind, it can be noted that there were a limited number of paper manufactures available during the time period. Therefore, the likelihood that a majority of the population purchased their paper goods from one of these manufacturers is significant. As a result, if one examines the number of documents which Cornwell claims to have done, matching watermarks will almost inevitably be found. It is also important to bear in mind that out of the hundreds of letters written, Cornwell matched only a few watermarks to those supposedly written by both Sickert and the Ripper. Furthermore, it has already been established that a majority of the Ripper letters were concluded to be hoaxes and their authenticity cannot be verified.

In the above example, Cornwell's theory of Sickert as the murderer may have clouded her perception of the material, causing her to take evidence which is subjective in nature and hold it as concrete.

Her persistence to her thesis enabled her to draw such conclusions and uncover the "evidence" she was perusing. Although Cornwell may have unveiled a connection between Sickert and *some* of the Ripper letters, the extent of the connection remains undetermined. In fact, her mitochondrial DNA and watermark evidence supports the idea that Sickert may have been responsible for submitting a hoax letter, but presents no conclusive evidence as to his involvement in the actual murders.

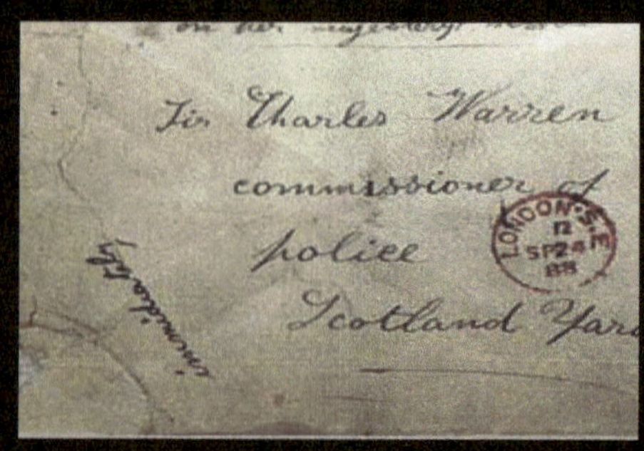

To Charles Warren
commissioner of
police
Scotland Yard

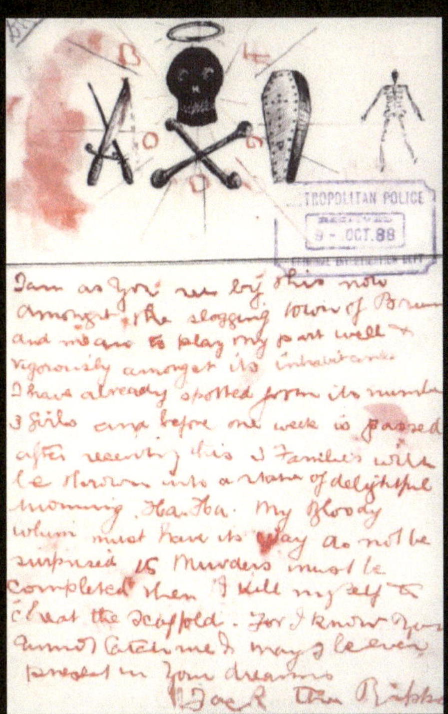

TROPOLITAN POLICE
9 - OCT.88

I am as you see by this now
Amongst the slogging town of Brum
and means to play my part well &
vigorously amongst its inhabitants
I have already spotted from its numbers
3 Girls [illegible] before one week is passed
after receiving this 3 Families will
be thrown into a state of delightful
mourning. Ha Ha. My bloody
whim must have its way so not be
surprised 15 Murders must be
completed then I kill my self to
cheat the scaffold. For I know you
cannot catch me & may I be ever
[illegible] in your dreams
Jack the Ripper

From the opening pages of her book, Cornwell calls particular attention to, and goes to great lengths to highlight Sickert's impotency and sexual inadequacies. She contributes the roots of his shortcomings to numerous surgical procedures aimed at correcting a fistula on his penis. A fistula is an abnormal connection between an organ, vessel, or intestine and another structure. In her work, Cornwell writes, "there are no references to fistulas of the penis in the medical records I consulted." More specifically, a review of Sickert's doctor's medical records failed to "unearth any mention of his threatening so-called fistulas of the penis" (Cornwell, 2002). Despite clearly stating that no record exists stating the exact nature of his fistula or why three surgeries were required to repair it, Cornwell adamantly contends that it was on his penis. Apparently, her conclusion is based solely on a brief conversation with Sickert's nephew, John Lessore.

She argues that the repercussions of this unverified anatomical anomaly fueled Sickert's hatred of women, thus causing him to commit the brutal series of Ripper murders. Although it is believed that Sickert did have a fistula, Cornwell's own research suggests that it was located on his anus or rectum. Cornwell determined that Dr. Alfred Duff Cooper of St. Mark's Hospital performed at least one of Sickert's surgeries. Dr. Cooper was known for performing surgeries on the vagina, anus and rectum, with no evidence suggesting he performed (or was qualified to perform) surgeries on the penis. Cornwell also supports her position that Sickert was impotent by stating he never fathered any children. However, she later contradicts herself by concluding his marriage to Ellen Cobden ended after she learned of his infidelities and conception of an illegitimate son (Cornwell, 2002). Despite numerous efforts, these researchers found no further evidence to support Cornwell's claim that Sickert was plagued by an extremely rare fistula on the penis which resulted in his impotency. Furthermore, no such evidence exists to suggest that such a deformity resulted in any subsequent murderous desires.

Cornwell's theory of Sickert as the murderer was also supported by her belief that his paintings portrayed similar features of the Ripper murders. According to Cornwell,

Sickert's paintings show morbidity, violence, and a hatred of women, while they also bear a resemblance to mortuary and crime scene photographs of Jack the Ripper's victims (Cornwell, 2002, p. 11). For example, in Sickert's painting, *Nuit d' Été*, or *Summer Night*, a nude woman is "grotesquely sprawled on an iron bedstead" (Cornwell, 2002, p. 115). At this point, Cornwell reminded the readers that Mary Ann Nichols was murdered on a summer night and suggested that the woman in the picture resembles Nichols (Cornwell, 2002, p. 115). She attempted to use the images portrayed in Sickert's artwork to support her thesis, claiming the similarities are too overwhelming to be done by anyone other than the Ripper. In reality, Cornwell only interpreted Sickert's painting as a type of confession that he was the Ripper, instead of realizing the painting could, in fact, portray something else that was meaningful to Sickert. There is no doubt that Sickert's art often portrayed violence and murder.

Several problems with this conclusion can be drawn. First of all, art is highly subjective to the viewer's interpretation and only the artist knows the entire truth behind its intended meaning. Cornwell claims to have uncovered clues left behind in Sickert's paintings which indicate his involvement in the murders. Again, the problem with such statements lies within the viewer's interpretation and presents one important question. How was Cornwell able to successfully decode Sickert's works of art and arrive at her conclusion? Sickert was a famous painter; his work has been studied by countless art critics for over a century, and no one has ever before managed to reveal such a connection. How then, did Cornwell, a fictional writer and ex-medical examiner's assistant (with no artistic expertise) manage to uncover this hidden secret and crack one of the biggest murder mysteries of all time?

Perhaps the answer to this complicated question is actually quite simple. During the time of the murders, an overwhelming amount of graphic detail surrounding the murders was published by the broadsheets and was readily available to the public. In fact, a quick search of news archives from the time quickly revealed a plethora of gruesome and detailed descriptions of the murders. If these researchers were able to uncover century old archives from newspapers such as the *London Times*, *Star Newspaper* and *Daily Telegraph*, there is no doubt that Sickert also had access to such details. After all, he did live in London at the time of the murders and was an avid reader. His talents as an artist would have easily enabled him to create paintings which depicted the murders in great detail. Therefore, any "evidence" that Sickert was the Ripper, based on his paintings, cannot be considered credible to solving the case. Moreover, Jack the Ripper made sure to not leave any evidence behind at the crime scene that could implicate him; therefore, it is highly unlikely that, if Sickert was the Ripper, he would portray images in his paintings that resembled the crimes of Jack the Ripper.

Yet another problem with Cornwell's premise is her complete rejection of the idea that the Ripper could have been a doctor, as had been theorized. Cornwell claimed that "surgical precision" is not required to remove organs from a body and that "it would be difficult for even a surgeon to "operate" when frenzied and in the dark" (Cornwell, 2002, p. 189). However, if it would have been difficult for a surgeon to remove an organ in the dark while "frenzied," wouldn't it be even more difficult for someone without any medical background or surgical skills to perfectly remove an organ and take it with him? Cornwell disregarded the conclusions made by Dr. Phillips, the police surgeon who examined the body of Annie Chapman. He thought that the injuries were made by someone who had "considerable anatomical knowledge and skill because the uterus had been removed by one who knew how to find it" (Wolf, 2008, p. 3344).Cornwell claimed that Sickert could have learned of the locations of the organs through the different medical books released at the time (Cornwell, 2002, p. 189). However, even if Sickert had read these books, we suggest that it still would have been difficult for him to locate and remove the organs without having undergone any medical training. Despite her best efforts, Cornwell's theory of the crimes does not convince the writers that Walter Sickert was Jack the Ripper. We urge the reading audience to read Cornwell's book for further and more detailed understanding of her premise.

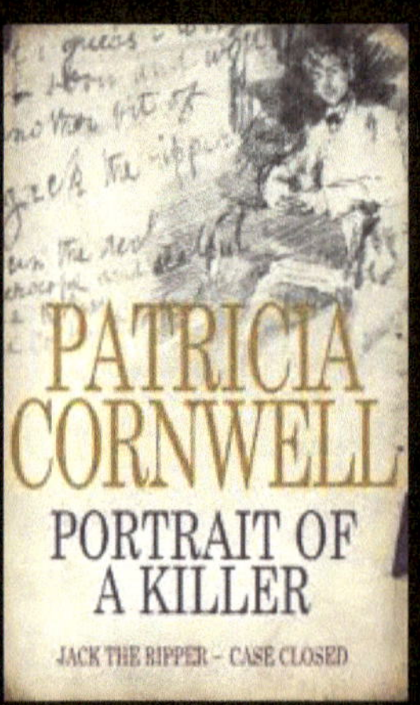


## Francis Tumblety: Jack the Ripper

The writers wish to suggest that a more viable suspect is Francis Tumblety, a well-traveled man of considerable wealth and questionable integrity, born in Ireland in 1833. He and his family immigrated to America and settled in Rochester, New York while he was still a boy (Begg & Evans, 2002). During his childhood, he was known to peddle pornographic literature through the canals of Rochester, NY (Curtis, 2001). One individual described him as "a dirty, awkward, ignorant, uncared-for, good-for-nothing boy, utterly devoid of education" (Begg & Evans, 2002). As an adult, Tumblety worked in a Rochester hospital whereby he "specialized in treating ailments "peculiar to women," performing hysterectomies, and treating the "problems of youths"" (Begg & Evans, 2002). Therefore, because of his prior experience with treating women, Tumblety had the medical knowledge that the Ripper seemed to possess as he was mutilating the women and removing their organs.

Later he spent time in the U.S., Canada, and England working as a physician. During the 1850's to 1860's, he was suspected of killing two women by issuing them lethal doses of medicine. He was also arrested in Canada "for attempting to perform an abortion on a 17-year-old prostitute;" however, his lawyers claimed that the detective and the prostitute framed him, which led to Tumblety's release from prison, "despite his reputation as an illegal abortionist" (Begg & Evans, 2002). In 1860, Tumblety was found guilty of manslaughter for the death of a patient under his care; consequently, he fled the area and was not seen there again (Begg & Evans, 2002).

During the Civil War, Tumblety established a medical practice in Washington, D.C. During this time, he threw a party in which no women were invited. When guests inquired about the lack of women at the party, "Tumblety launched into a homily against womankind, "especially fallen women." He then showed his guests a collection of anatomical specimens, which comprised about a dozen jars containing the wombs (including female uteri) "of every class of women"" (Begg& Evans, 2002). After being arrested and later exonerated in connection with the Lincoln assassination, he set sail for London in 1860.

Tumblety's hatred of women may have originated from his first and only marriage. He married an older woman whom he deeply loved; however, she was, by account, flirtatious and, as he later discovered, also a prostitute who was seeing other men. After this betrayal by his wife, he gave up on all women (Begg& Evans, 2002). In fact, "many people who knew him believed he disliked and avoided women" (Douglas &Olshaker, 2000, p. 70). Therefore, it is clear that Tumblety had an intense hatred for women, especially prostitutes, which could have motivated him to violently kill the five prostitutes. Furthermore, with his medical knowledge and past specialty in treating ailments common to women, Tumblety easily had the capabilities to identify the parts of the female reproductive system, which showed in his collection of female parts.

In June of 1888, he traveled to Liverpool where he was arrested several times for a variety of offenses including gross indecency and assault against his alleged homosexual partners. On November 7, 1888, he was arrested for "gross indecency with another man," which was a misdemeanor; therefore, he was released from prison within twenty-four hours (Begg& Evans, 2002). Two days after Tumblety's release from prison, Mary Jane Kelly was murdered, on November 9, 1888 (Begg& Evans, 2002). Thus, as supported by the facts, Tumblety arrived in London shortly before the murders began and was seemingly available to kill the prostitutes, as shown in his release from prison two days prior to the last Ripper murder.

He was then arrested on suspicion of committing the Ripper murders, but in December of 1888, he fled to New York City where he was placed under police surveillance. It is also reported that a team of inspectors from England followed Tumblety to New York Auckland, 1889).

After managing to lose his police detail, and as a result of the English authorities having no further evidence to hold against him, the investigation into Tumblety went cold. He later told a New York reporter that he did indeed frequent the Whitechapel area. In fact, Tumblety stated:

> *"Someone said that Jack the Ripper was an American, and everybody believed the statement. Then it was the universal belief among the lower classes that all Americans wear slouch hats; therefore, Jack the Ripper must wear a slouch hat. Now, I happen to have a slouch hat, and this, together with the fact that I was an American was enough for the police. It established my guilt beyond any question" (Daily Argus News, 1889). He was also asked about his supposed status as a women hater, to which he replied with laughter and stated, "I don't care to talk about the ladies" (Daily Argus News, 1889).*

Interestingly, when Tumblety left London, the Ripper murders ceased. Also, his "demonstrated and deep-seated hatred of women" would have been a motivating factor in his decision to target women, particularly prostitutes (Gordon, 2003, p. 8).

Furthermore, Tumblety presented himself as the best Jack the Ripper suspect because of his extensive medical knowledge and training. The belief that the Ripper was a doctor originated from the Ripper's method of killing and mutilating the victims. "In fact, with the exception of Elizabeth Stride, each of the Ripper victims' mutilations revealed signs of medical training" (Gibson, 2004, p. 148). Moreover, the physician who examined the body of Polly Nichols, Dr. Rees Ralph Llewellyn, stated that, "The murderer must have had some rough anatomical knowledge. He seems to have attacked all the vital parts" (Gibson, 2004, p. 148).

Additionally, the coroner involved in the Annie Chapman murder, Wynne Baxter, stated that "the injuries had been made by someone who had considerable anatomical skills and knowledge...The organ had been taken by one who knew where to find it...No unskilled person could have known where to find it or have recognized it when it was found" (Gibson, 2004, p. 148). Tumblety had the necessary medical skills and experience to kill the victims and to identify and remove their organs from their bodies. Based on Tumblety's medical knowledge as well as his hatred for women and prostitutes, we propose that he was most likely to have been Jack the Ripper, although we believe the Ripper's identity will always be in debate .

## ABOUT THE AUTHORS:

**Kathleen Gahagan** — **Gahagan** earned both her Masters and Bachelor of Science in Criminal Justice with honors from West Chester University. She completed her internship with the Upper Dublin Township, PA Police Department in the summer of 2010. She will be attending Temple University Municipal Police Academy beginning in May 2012. Her goal is to use the skills she has acquired to begin a career in law enforcement.

**Ross Neely** — Neely received his Master's degree in Criminal Justice from West Chester University of Pennsylvania. After graduating from West Chester University in 2010 with a degree in Criminal Justice, he began working as Adult Probation and Parole Officer. He plans to further pursue a career in the criminal justice field by working in federal law enforcement .

## REFERENCES

Begg, P., & Evans, S. (2002). Jack the Ripper: Two suspects 'on Trial'. *British Heritage, 32*(6).

Bell, D. (1974). Jack the ripper: The final solution. *Criminologist*, 9 (1).

Cohen, D. (2002). *Jack the ripper: Theories and misconceptions.* Retrieved from http://www.slate.com/id/207350/

Cornwell, P. (2002). *Portrait of a killer: Jack the Ripper case closed.* New York, New York: The Berkley Publishing Group.

Cullen, T. A. (1965). *When London walked in terror.* Boston, Massachusetts: Houghton Mifflin Company.

Curtis, P.L. (2001). *Jack the ripper and the London press.* New Haven, CT: Yale University Press.

Daily Argus News. (1889, February 26). They suspected his hat: why Dr. Tumblety was arrested. *The Daily Argus News.*

Douglas, J., &Olshaker, M. (2000). *The cases that haunt us: From Jack the Ripper to JonBenet Ramsey, the FBI's legendary mindhunter sheds light on the mysteries that won't go away.* New York, New York: Scribner.

Eckert, W.G. (1981). The whitechapel murders: The case of jack the ripper. *American Journal of Forensic Medicine & Pathology*, 2 (1).

Evans, S, & Gainey, P. (1998). *Jack the ripper: first American serial killer.* New York, NY: Kodansha America.

Eyres, J (Director). (2000). *Ripper: Letter from Hell* [VHS].

Foran, J. (2006). The evil deeds of Dr. Cream. *Beaver, 86*(4).

Fox, J.A., & Levin, J. (2005). *Extreme killing: Understanding serial and mass murder.* Thousand Oaks, CA: Sage Publications Inc.

Giannangelo, S.J. (1996). *The psychopathology of serial murder: A theory of violence.* Westport, CT: Praeger Publishers.

Gibson, D. C. (2004). *Clues from killers: Serial murder and crime scene messages.* Westport, Connecticut: Praeger Publishers.

Goldman, D. (2001). The monster of Whitechapel: The continuing mystery of Jack the Ripper. *Biography Magazine, 5*(10), 24.

Gordon, R. M. (2003). *The American murders of Jack the Ripper.* Westport, Connecticut: Praeger Publishers.

Holmes, R.M, & DeBurger, J. (1988). *Serial murder: Studies in crime, law, and criminal justice.* Thousand Oaks, CA: Sage Publications, Inc.

Holmes, R. M., & Holmes, S. T. (2002). *Profiling violent crimes: An investigative tool* (3rd ed.). Thousand Oaks, California: Sage Publications, Inc.

Keppel, R. D., Weis, J. G., Brown, K. M., & Welch, K. (2005). The Jack the Ripper murders: A modus operandi and signature analysis of the 1888-1891 Whitechapel murders. *Journal of Investigative Psychology and Offender Profiling, 2*(1), 1-21. Princeton, NJ: Princeton University Press.

Leach, K. (1999). *In the shadow of the dream child: A new understanding of Lewis Carroll.* Holland Park, London: Peter Owen Publishers.

Morley, C.J. (2005). *Jack the ripper: A suspect guide.* New York, NY: Random house.

Murray, A. (2004). Jack the Ripper, the dialectic of enlightenment and the search for spiritual Deliver ance in White Chappell, Scarlet Tracings. *Critical Survey, 16*(1), 52-66.

Ramsland, K. (2006). *Inside the minds of serial killers: Why they kill.* Westport, Connecticut: Praeger Publishers.

Rosenhek, J. (2005). *The madman of Mcgill.* Retrieved from http://www.doctorsreview.com/history/jun05_history

Rubinstein, W. D. (2000). The hunt for Jack the Ripper. *History Today, 50*(5).

Schmid, D. (2005). *Natural born celebrities: Serial killers in American culture.* Chicago, Illinois: The University of Chicago Press.

Stowell, T.D. (1970). Jack the ripper: A solution? *The Criminologist*, 5.

Turvey, B.E. (1999). *Criminal profiling: An introduction to behavioral evidence analysis.* San Diego, CA: Academic Press.

Wallace, R. (1996). *Jack the ripper: Light-hearted friend.* Melrose, MA: Gemini Press.

Wolf, G. (2008). A kidney from hell? A nephrological view of the Whitechapel murders in 1888. *Nephrology Dialysis Transplantation, 23*(10), 3343-3349.

**PHOTO CREDITS:**

Cover photo: Woody

Probeaway– Life Hacks internet

Little White Chapel, Blog About History

Jack the Ripper 1888—Elizabeth Stride

Mary Ann Nichols—History.comSpooky Isles, Jack the Ripper Victimes

The Pennington Edition—WhiteChapel Street

Street Spirit, a publication of the American Friends Service Committee

Fi2X Entertainment

Murder by Gaslight, May 1, 2010

On the track of Jack the Ripper, by Alain, TRE-KEARTH (Learning about the world through photography

That Videogame Blog, Oct. 7th, 2009

http://www.casebook.org/victims/stride.html

Tough Cases.net, Jack the Ripper Case file

Jack the Ripper, Whitechapel Murders, Paperless Archives .com

Spons, Screens, Jack the Ripper: Letters from Hell PC

WD FYFE, Jack the Ripper: The Last of its Kind

Art Blog by Bob, Late Bloomer, Oct. 24, 2008

Wikipedia, the free Encyclopedia

The Telegraph, Scotland Yard Fights to keep Jack the Ripper files secret. May 15, 2011, by David Barrett

Ghost Theory

EDP 24, June 19, 2012. Jack the Ripper Murdered 11 Women, evidence now suggests, March 8, 2008

Examiner - Society and Culture, Dan Norder

Crime History 101, Who was the First American serial Killer, August 23, 2009

The Daily Telegraph Australia, May 9, 2012

The Telegraph, Jack the Ripper: The Suspects, June 11, 2012

Pulp International, Sept. 30, 2009, From Hell

WhiteChapel Jack, The Legend of Jack the Ripper

Socyberty, Jack the Ripper Victims by Kim Seabrook in History, Oct. 21, 2009

Cult, Movie Forums, "So Sweet So Dead" (Roberto Montero, 1972), July 2008.

# SPECIAL FEATURE: FORENSIC DISCIPLINES

*As compiled by the Editorial Staff*

FORENSIC ACCOUNTING

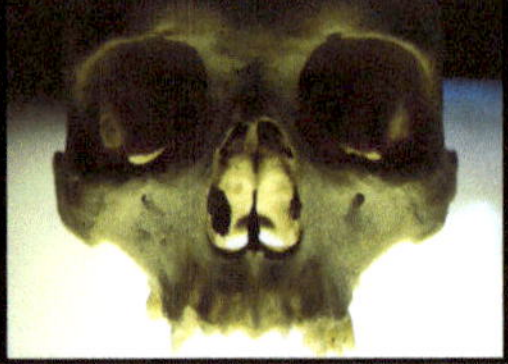

FORENSIC ANTHROPOLOGY

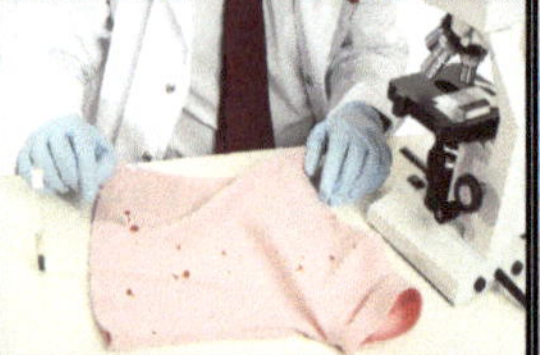

FORENSIC BIOLOGY

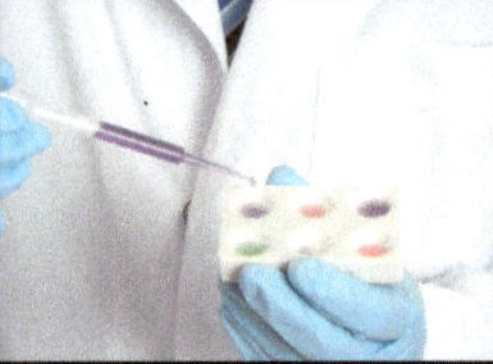

FORENSIC CHEMISTRY

FORENSIC ENTOMOLOGY

## *What is a Forensic Accountant?*

Forensic Accountants investigate crimes within the field of finance, like fraud.
They are also know as fraud investigators, investigative accountants, forensic auditors or fraud auditors.

They conduct audits by actively looking for signs of fraud, examining financial statements for accuracy and completeness, seeking out internal databases and court records.

They meet with a government representative, attorney or other clients to learn the specifics of the alleged fraud and conduct interviews with the accused and other persons associated with the investigation.

They search records such as bank statements, credit statements, journal, ledgers, databases, emails, memos, letters, and other relevant documents that will offer information into the financial situation.

They are responsible for identifying and giving interpretations based on the collected information, and may be asked to testify as an expert witness during trials and hearings.

## *What is a Forensic Anthropologist*

Forensic anthropologists are often self-employed or act as consultants to the police or industry in addition to their full-time jobs. Many pursue employment at a college university in teaching or research and gather experience while working.
They arrive at the scene of the crime to assist in locating and recovering suspicious remains.

They are commonly called upon to suggest the species, age, sex, race, stature, weight and unique features of a decedent from the skeleton.

They may also be asked to analyze the remains to determine if there is any evidence of trauma, the manner and cause of death.

## *What is a Forensic Biologist?*

They very carefully and thoroughly examine biological fluids (blood, saliva, urine, feces, seminal and vaginal fluid] found on clothing, weapons, and other evidence recovered at a crime scene by using a variety of procedures and testing methods.
Samples suspected of being biological fluid are often analyzed for DNA in order to determine the probability of it being related to or originating from a suspected individual involved in the case.

They are responsible for identifying and giving interpretations based on the collected substances, and may be asked to testify as an expert witness during trials and hearings.

## *What is a Forensic Chemist?*

Forensic Chemists apply their knowledge of Biology, Genetics, and Chemistry to analyze specimens using the disciplines of Criminology and Toxicology.

They very carefully and thoroughly examine items suspected of being drugs (pills, powders, and plant material] and determine the identity of these items by performing a variety of tests.

They are involved in the analysis of chemical substances involved in arson, explosions and other fire investigations.

They examine trace evidence such as hair, fiber, glass, soil and paint to determine its origin and association with suspected individuals involved in the case.
They are responsible for identifying and giving interpretations based on the collected substances, and may be asked to testify as an expert witness during trials and hearings.

## *What is a Forensic Entomologist*

Forensic entomologist are often self-employed or act as consultants to the police or industry in addition to their full-time jobs. Many pursue employment at a college university in teaching or research and gather experience while working.

They are commonly called upon to use insects to determine the circumstances of an abuse or rape, to determine the circumstances of crime before and after death, or to determine the time of death in homicide investigations.

They arrive at the scene of a crime to collect insects from a body, to collect insects present after body removal and to carefully package and ship collected insects for proper analysis.

They can potentially use DNA technology to help determine insect species, to determine a relationship between perpetrator and suspect and they can also

FORENSIC LINGUISTIC

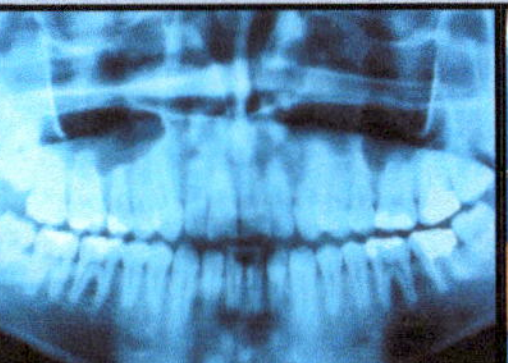
FORENSIC ODONTOLOGY

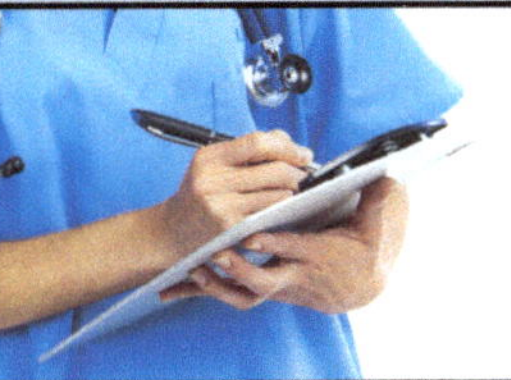
FORENSIC NURSING

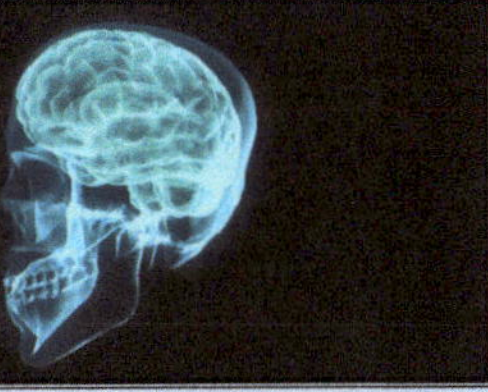
FORENSIC PSYCHOLOGY

FORENSIC TOXICOLOGY

determine insect species, to determine a relationship between perpetrator and suspect and they can also assist in toxicological analysis by using insects recovered from human remains, often where tissues needed for toxicological analysis disappear.

Urban entemologists study insects such as termites, fleas and cockroaches that can cause damage to structure or infest areas. Some specialize in stored product entomology, which deals with intentional or unintentional insect contamination of food and beverages.

Medico legal entemology is a specialization in which insect evidence can determine things such as where a crime took place, or if death occurred by insect.

They may be asked to act as an expert witnesses in criminal and civil proceedings dealing with the evidence they've collected and analyzed.

## What is a Forensic Linguist?

Forensic linguists often pursue academic careers, teaching and conducting research, as well as work as private consultants providing their services to law enforcement agencies to help solve crimes and to legal professionals to provide expert testimony on their findings in the case.

Their expertise is utilized in areas such as voice identification, author identification, in regards to confessions, ransom letters, plagiarism, word meaning, contract disputes, fraud, perjury, legal language, document validation, reconstruction of telephone conversation or text message exchanges, all for purposes of identifying a suspect or absolving someone that was falsely accused.

## What is a Forensic Odontologist?

Typically, a forensic odontologist works full time as a dentist and is only utilized as needed for a case.

They use teeth as a source of identification for a crime, identify bite marks on a victim or an item of evidence, compare bite marks with the teeth of a suspect and identify unknown bodies through dental records.

They are responsible for identifying and giving interpretations based on the collected information, and may be asked to testify as an expert witness during trials and hearings.

## What is a Forensic Nurse?

A registered nurse that specializes in forensics that provides specialized care for patients, for both victims and perpetrators of violence, such as physical, psychological and social trauma that occurs during an assault or abuse.

They are skilled in injury identification and evaluation, documentation of the patients incident, collection and proper storage of biological and physical evidence.

They work with sexual assault patients, patients in a correction institution or a psychiatric hospital.

They are trained to work with domestic violence, child abuse and neglect, death investigation, elder mistreatment, correction, emergency services, mental health and public health situations.

They are responsible for giving interpretations based on the collected information, and may be asked to testify as an expert witness during trials and hearings.

## What is a Forensic Psychologist?

Forensic psychologists are self-employed or act as consultants to the police or industry in addition to their full-time jobs as licensed clinical psychologists.

Many pursue a career at a college university in teaching or researching human behavior, criminology, and the legal system and gather experience while working.

They can specialize in the areas of family, civil or criminal court, acting as an expert witness and are called upon to assist in a wide variety of legal matters, including the mental state of criminal defendants (insanity, competency, etc.), jury selection, child custody/family law, violence risk prediction, mediation/dispute resolution, discrimination, civil damages, social science research (e.g. recovered memory) and civil commitment.

## What is a Forensic Toxicologist?

They use toxicology and other disciplines such as analytical chemistry, pharmacology and clinical chemistry to aid medical and legal investigations of death, poisoning and drug use.

Testing is done on a variety of physiological samples such as blood, urine, oral fluid and tissues to thoroughly investigate exposure, ingestion, abuse and the presence of drugs and/or other toxic substances.

They are responsible for giving interpretations based on the collected substances, and may e asked to testify as an expert witness during trials and hearings.

# Police and PTSD:

## A Personal Journey from the Field

By Craig Rickard

By Craig Rickard

**Shots Fired." "Officer Down."**
**"I've been hurt in an accident."**
**"I have a dead body." "I need help.........."**

As a former police chief in modern America, I know first-hand that the quotes above are often used by police officers upon arrival at a critical incident. I have used similar phrases and terms in emergencies myself.

Police officers are generally the very first responders to arrive at a scene.

*First responders,* in the medical, forensic and law enforcement fields, are those individuals trained to respond to the worst emergency scenes that society renders. It could be said the defining critical incident for first responders was experienced on American soil on September 11, 2001.

**A police officer's daily job is to perform routine patrol, look for trouble or emergencies or be ultimately dispatched to such incidents. Many of the calls responded to throughout an officer's career can be classified as *critical incidents.***

**Officer.com (1),** one of the better police support web sites, defines a ***critical incident*** as follows:

*Any event that has a stressful impact sufficient enough to overwhelm the usually effective coping skills of an individual. Critical incidents are abrupt, powerful events that fall outside the range of ordinary human experiences. These events can have a strong emotional impact, even on the most experienced officer or deputy.*

Quite often, police officers and supervisors alike can experience significant emotional and psychological effects from these incidents.

These effects sometimes meet the criteria for **Post-Traumatic Stress Disorder (hereafter called PTSD)**.

The **Diagnostic and Statistical Manual of Mental Disorders (DSM-IV)** defines **Post-Traumatic Stress Disorder**:

*A severe anxiety disorder that can develop after exposure to any event that results in psychological trauma. This event may involve the threat of death to oneself or to someone else, or to one's own or someone else's physical, sexual, or psychological integrity, overwhelming the individual's ability to cope. (2)*

It has become more important than ever for modern police officers to recognize their responses to stress, what to do with those responses, and the appropriate methods of verbalizing their feelings. Subsequent to a critical incident, officers may experience a wide range of emotions, including fear, anxiety and confusion.

Police chiefs throughout the country are keenly aware of these after effects. What is seen by police personnel

after arriving at such incidents is sometimes horrific and, at times, haunting. The informed supervisor should be aware of these and other typical symptoms of PTSD, and be educated in how to help the victim officer find the means to address his/her emotions and memories. It should be noted that not all officers who are subjected to acute forms of critical incidents experience PTSD. However, after a critical incident, fellow officers, supervisors, and especially family members, may begin to see a significant change in the behavior of a sufferer of PTSD.

I know of which I speak. In **1985,** I was on normal patrol as a police officer, never expecting that I would be compelled to use deadly force against a drug dealer. The individual had returned home in time to see federal officers performing a drug raid at his meth lab apartment. After a lengthy pursuit through two suburban towns, the actor crashed thru a police roadblock, injuring a citizen and disabling a patrol unit. When the subject bottomed out in the ice and snow, I approached. He somehow got his sports car running again and continually attempted to run me over, before I could get back to my patrol vehicle. I was completely exposed and vulnerable. As a result, I shot once, injuring him.

While other officers were securing him, I discovered a 17 year old prostitute curled up on the floorboard of his vehicle. The radio BOLO (Be On Lookout) reported only, "one occupant." This haunted me for many years to come, as she could have easily been killed. I now have two daughters of my own and the child's helplessness that night still makes me cringe when I allow myself to think about it.

The answer to my 1985 post incident shakiness was for my fellow officers to take me out drinking after the shift was over. It became customary for many of us to handle critical incidents in much the same way...drink it away and drink to forget. We all know that this does not work, and leads only to poor consequences and non-resolution of the unresolved memories and feelings.

**Diagnostic Criteria of Post - Traumatic Stress Disorder**

**Diagnostic symptoms** for PTSD include, but are not limited to re-experiencing the original trauma(s) through flashbacks, nightmares, avoidance of stimuli associated with the trauma, increased arousal –such as difficulty falling asleep or staying asleep, anger, and hyper vigilance. I personally experienced *all* of these.

**Formal diagnostic** criteria (both from **DSM-IV-TR** and **International Statistical Classification of Diseases and Related Health Problems, or ICD-10**) require that the symptoms last more than one month and cause significant impairment in social, occupational, or other important areas of functioning.

## Symptoms

People with PTSD relive the event over and over again. They may experience nightmares, sad reminders or flashbacks.

Symptoms of PTSD fall into three main categories:

1. **Repeated "reliving" of the event, which disturbs day-to-day activity**

- Flashback episodes, where the event seems to be happening again and again
- Recurrent distressing memories of the event
- Repeated dreams of the event

Physical reactions to situations that remind you of the traumatic event

### 2. Avoidance

- Emotional "numbing," or feeling as though you don't care about anything
- Feelings of detachment
- Inability to remember important aspects of the trauma
- Lack of interest in normal activities
- Less expression of moods
- Staying away from places, people, or objects that remind you of the event
- Sense of having no future

### 3. Arousal

- Difficulty concentrating
- Exaggerated response to things that startle you
- Excess awareness (hyper vigilance)
- Irritability or outbursts of anger
- Sleeping difficulties

The PTSD sufferer might also feel a sense of guilt about the event (including "survivor guilt"), and the following symptoms, which are typical of anxiety, stress, and tension:

- Agitation, or excitability
- Dizziness
- Fainting

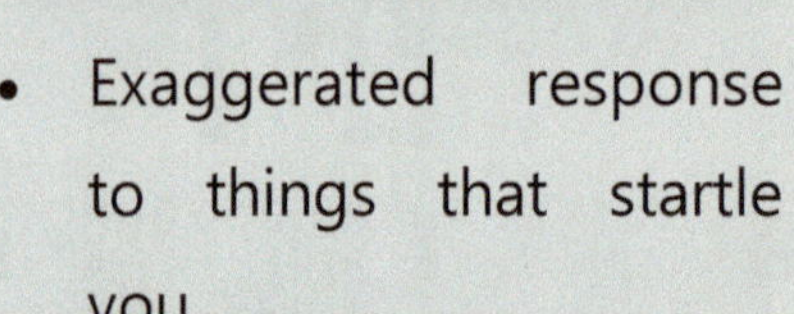

- Exaggerated response to things that startle you
- Excess awareness (hyper vigilance)
- Irritability or outbursts of anger
- Sleeping difficulties

The PTSD sufferer might also feel a sense of guilt about the event (including "survivor guilt"), and the following symptoms, which are typical of anxiety, stress, and tension:

- Agitation, or excitability
- Dizziness
- Fainting
- Feeling your heart beat in your chest (palpitations)
- Fever
- Headache
- Paleness

## Signs and tests for PTSD

Unfortunately, there are no tests that can be done to diagnose PTSD. The diagnosis is made based on a certain set of symptoms that continue after someone has experienced extreme trauma. In order to rule out other illnesses, a doctor would need to do psychiatric and physical exams (3&5).

## Supervisors and Managers

**In my own career** I have experienced PTSD. I have witnessed other officers who were surely sufferers of PTSD and did not know it. When I began serving in law enforcement in **1977,** and for many of the years thereafter, **the term PTSD was not really used. *"Shell shock", "battle fatigue",* etc., were common terms that referred to people/soldiers, who had seen way too much and were having difficulty dealing with their prob-**

**lems. Their emotional or psychiatric reactions to their specific horror or other traumatic exposure rendered them stigmatized as weak and unable to cope. They were sometimes whispered about.**

**2012. Policing in America** has changed dramatically from when I started. Bullet proof vests, OC Spray, multi-channel radios, semi and automatic weapons, Taser weapons, less-lethal weapons, etc., were not even considerations for the officer on patrol. **Today,** no officer would survive without them.

**Terrorism,** both domestic and abroad also makes it incumbent upon chiefs, supervisors, fellow officers and family members, to become educated and adept at identifying the warning signs of PTSD. Officer suicides, call-outs, job loss, panic attacks, alcoholism, addictions of every kind and more, can be dramatically reduced through open discussion, education and training.

Support groups are available to all. CISM or Critical Incident Stress Management should be part of every police academy and ongoing annual certification. Peer counselors and department chaplains can enable the officer to vent therapeutically, without being seen as "weak" or "unable to handle the job". This stereotypical thinking has kept many officers from coming forward with their problems. The culture in law enforcement is sometimes very rigid and overly competitive. It is our job as leaders to ensure that our valued employees are cared for, trained and counseled in the most dangerous aspects of the job-***critical incidents.***

**Why** is it that we supply our officers with the best of equipment to go into battle with the criminal element, yet we fail to adequately train them skillfully, in the art of coping?

**In conclusion, police officers are killing themselves at a rate more than twice as high as the general public (McNamara, 1996). This in and of itself should lend the experts and individual police departments to devote training and prevention in the area of suicide to all officers (4).**

## Bibliography

Kulbarsh, Pamela. "Critical Incident Stress." Officer.com. N.P., 5 Oct. 2007. Web. 6 July 2012. <http://www.officer.com/article/10249385/critical-incident-stress>.

Diagnostic and Statistical Manual of Mental Disorders: DSM-IV4th edition. Published in Washington, D.C. by the American Psychiatric Association in 1994.

http://www.nimh.nih.gov/health/publications/post-traumatic-stress-disorder-ptsd/what-are-the-symptoms-of-ptsd.shtml

http://www.tearsofacop.com/police/articles/dugdale.html

International Statistical Classification of Diseases and Related Health Problems: ICD-10th Revision Edition. Geneva: World Health Organization, 1992. Print.

## ABOUT THE AUTHOR

**Craig Rickard's** law enforcement career began in 1977 in Pennsylvania at the Montgomery County Sheriffs Department and culminated in his retirement as Chief of Police at the Jenkintown, PA Police Department. Craig rose through the ranks first as a Certified Crime Prevention Officer, then as a Sergeant in charge of Criminal Investigations. As a Criminal Investigator,he received extensive FBI and investigative training. He was a Class Representative at the FBI National Academy's 180th Session in 1995, wherein his major course of study was in the Interview and Interrogation Program and the Behavioral Science Unit. After retirement, Craig entered the private security sector and served as a regional manager for the Mid- Atlantic region, remaining in that capacity for six more years. He is married, has two daughters in college and runs his own home improvement business.

## PHOTO CREDITS

- Philippi Trust S. Af. Post Traumatic Disorder, www.philipitrust.co.za, May 5, 2011.
- NRI Neurologic Rehabilitation Institute at Brookhaven Hospital—Traumatic Brain Injury Maybe Linked to PTSD, March 20, 2012.
- Neuroanthropology, Cultural Aspects of PTSD: Thinking on the Meaning and Risk: Posted by lpfinley on June 4, 2008.
- University of Michigan Student Affairs.
- NY Times, April 4, 2009, Gunman kills 3 Police Officers in Pittsburg.
- Law Enforcement Today, *The Forgotten Element,* by Julie Adcock, *Zero Recidivism Rate*?, by Anne Bremer, *Can experts Spot Lawbreakers Early?,* by Jean Reynolds, *Wives On Duty—Listening* by Chaplain Alison P. Uribe, *Monday's Perspective* by July Adcock, *What Is (or Isn't a Crime?)* by Jean Reynolds, *Memorial Day* by Mark St. Hilaire, *Along Way to Run for Charity* by Jean Reynolds, *Swatter's Rights* by James P. Gaffney, *BCops Study Indicates Police Work May Have Adverse Health Effects,* by St. Hilaire.
- The Temple News, September 23, 2008. *That Missing Smile* by Kenneth Wise.
- Elegy for a Pig by Anthony Pizighelli.
- A Debt Repaid by Kenneth Wise.
- Public Safety Dispatchers: Disputes and Distress
- by Mark St. Hilaire.
- Training, friends and Eating Challenge.
- Line of Duty Officer
- Ramblings, July 18, 2010
- NeuroAnthraopology, Cultural Aspects of PTSD: Thinking on Meaning and Risk, June 4, 2008.
- 911 First Responders Health issues. Increased Risk of Heart Diseases, November 16, 2011
- Silverchips Online, 7/25/12, courtesy of National Geographic.com
- The day of that day: taking a Human Life.
- Predictive Policing
- Police Agencies Gear Up for NEW Year's Eve Gunfire.
- Officer Shot in Nassau County, NY.
- Scientific American: Soldiers' Stress: What Doctors Get Wrong About PTSD by David Dobbs, April 13, 2009.
- NY Times, April 4, 2009, Gunman Kills 3 Police Officers in Pittsburg.

## CALL FOR PAPERS AND ARTICLES

The International Academy of Forensic Professionals (IAFP) has issued an open-call for manuscripts, research reports, and other material suitable for all forensic science disciplines. Your professional offering should be submitted for review by the Editorial Board who will then determine appropriateness for online publication in our E-journal, *The Forensic Digest.*

All materials submitted must be original and not previously published in any other copyrighted works. Authors of accepted work are required to sign an exclusive copyright release for any manuscript submitted. The address for requesting further information or to submit your work is: FD@tiafp.org

## ARTICLE SUBMISSION GUIDELINES FOR THE FORENSIC DIGEST

*The Forensic Digest* welcomes first time writers as well as published authors. Writing is a valuable medium through which we share knowledge, opinion and research with others. Writing may be considered a professional obligation as it serves as a method of mentoring through the presentation of information that stimulates thoughts and ideas in the reader.

The editorial staff of Forensic Digest is aware that submission guidelines may often prove a stumbling block that prohibit rather than encourage writers. To that end, we have tried to make the process of article submission as "user-friendly" as possible. Please review the guidelines below and start writing!

**Manuscript submission guidelines:**

1. Submit your work in MS Word (.doc; .docx) or Rich-text Format (.rtf) only
2. Use Times Roman 12 point font
3. Double space your work
4. Develop a page one that includes:

   a. The working title of the manuscript
   b. The names of all writers the primary author is listed first, with the others to follow.
   c. Identify the educational degrees, if applicable, after each name.
   d. The primary contact source for article review, questions and comments.
   e. Contact information should include both a physical as well as an email address.

***Additional points:***

Conceptualize your finished manuscript at a minimum of four ages to no more than 12 pages in length, excluding page one. Graphs, tables etc. may be inserted in the manuscript or included at the end as appendices. All graphics must have a descriptive caption. If you are including any work not your own, you must have a statement of copyright permission from the copyright holder. Permission may be in the form of a formal letter or an email from the copyright holder. All citations and references must be formatted using the format found in the Publication Manual of the American Psychological Association (5th ed.), commonly referred to as APA guidelines.

Authors submitting to the *Forensic Digest* must certify in writing, at the time of their submission, that the article is not currently under review elsewhere for publication and that it will not be submitted elsewhere prior to editorial decision at the *Digest.* Certification may be made easily through an email to the editor. The primary author may write on behalf of all contributing authors to the article.

***And a few more points:***

- The editorial staff reserves the right to edit the material without changing the meaning or intent of the submission. All edits are returned to the (primary) writer for review and consideration.
- If your manuscript has been approved for inclusion in the *Forensic Digest*, you will be notified via email. At that time, you will be asked to submit a photograph and short biographical sketch.
- The editorial staff will select photos and graphics that best illustrate the intent and meaning of your article. We will send you the completed version of the manuscript before publication. It is your responsibility to address any concerns you might have at that time.
- Once your manuscript has been accepted for publication by *The Forensic Digest*, it becomes our property and we hold the copyright to it.
- All manuscripts should be submitted to: editorFD@tiafp.org

## Chateau Battiste Assisted Living

## Memory Care Community

# Chateau Battiste

## Assisted Living and Memory Care Community

SITUATED NEAR HEMET VALLEY MEDICAL CENTER
AND OTHER HEALTHCARE PROFESSIONALS READILY AVAILABLE
FOR SERVICES TO OUR RESIDENTS

Chateau Battiste at Hemet Street
161 North Hemet Street
Hemet, CA 92544

Phone: 951.927.6817
Fax: 951.927.8439

161 North Hemet Street

www.chateaubattiste.com
info@amrn.com

*Owned and Operated by a Registered Nurse*

## SEXUAL ASSAULT EXAMINATION TRAINING TOOLS

Set Includes:

- Sexual Assault Examination DVD and Instructional Reference Guide CD
- Photographic Slides of Adult Genital and Anal Injuries
- Photographic Slides of Child Genital and Anal Injuries
- Sexual Assault Examination Procedures Manual CD

**Total Value $419.90 Save $75.00 when you purchase the complete Tool Set: $344.90 ($9.95 S&H)**

## THE SEXUAL ASSAULT EXAMINATION: ESSENTIAL FORENSIC TECHNIQUES AND INSTRUCTIONAL REFERENCE GUIDE

This one hour DVD/video The Sexual Assault Examination: Essential Forensic Techniques depicts a visual re-enactment of the crime of sexual assault and the methodology used in the professional collection of forensic evidence by the Sexual Assault Forensic Nurse Examiner. This presentation is an indispensable tool for physicians, nurses, victim advocates, law enforcement officials and judiciary involved in the examination of victims of sexual violence, collection of evidence and prosecution of the perpetrator.

Topics covered are the role of the sexual assault team members, interview process, medico-legal examination, colposcopic photography, toluidine dye as used in sexual assault, victim follow-up care, evidence chain of custody and documentation. The accompanying CD-ROM Instructional Reference Guide is designed to be used in conjunction with the DVD.

## SEXUAL ASSAULT RESPONSE TEAM PROCEDURES MANUAL

This manual is designed to serve three purposes: to be used as a guideline for general and practical information for the sexual assault response team (SART); as a framework for training; and as a practical reference tool. It contains policies, step-by-step procedure guides, reference materials, check lists, medical terminology, work sheets, diagrams, data collection logs and other sample forms necessary for the accurate documentation of forensic evidence collection.
**Cost: $175.00 ($9.95 S&H)**

## PHOTOGRAPHIC SLIDES OF ADULT GENITAL AND ANAL INJURIES

These adult images are an excellent resource for teaching and research needs. Carefully selected, the genital and anal findings, photographed with the colposcope, clearly identify typical trauma relating to sexual violence. **50 CD IMAGES Cost: $59.95 ($7.95 S&H)**

## PHOTOGRAPHIC SLIDES OF CHILD GENITAL AND ANAL INJURIES

These child images are an excellent resource for teaching and research needs. Carefully selected, the genital and anal findings, photographed with the colposcope, clearly identify typical trauma relating to sexual violence. **50 CD IMAGES Cost: $59.95 ($7.95 S&H)**

If you have problems purchasing products,
please contact us: info@taife.com or telephone 760.322.9925.

www.ingramcontent.com/pod-product-compliance
Lightning Source LLC
Chambersburg PA
CBHW051020180526
45172CB00002B/423

* 9 7 8 1 3 0 0 0 9 7 2 0 4 *